AMERICA'S PUBLIC LANDS

Series Editor

Char Miller
POMONA COLLEGE

Advisory Board

Douglas Brinkley
RICE UNIVERSITY

Jackie Gonzales
HISTORICAL RESEARCH ASSOCIATES INC.

Patricia Nelson Limerick
UNIVERSITY OF COLORADO BOULDER

THE BEARS OF GRAND TETON

A Natural and Cultural History

SUE CONSOLO-MURPHY

UNIVERSITY OF NEBRASKA PRESS

Lincoln

Huyler, Jack. *And That's the Way It Was in Jackson Hole, Second Edition*. Jackson Hole Historical Society and Museum, 2003. Used by permission.

The University of Nebraska Press is part of a land-grant institution with campuses and programs on the past, present, and future homelands of the Pawnee, Ponca, Otoe-Missouria, Omaha, Dakota, Lakota, Kaw, Cheyenne, and Arapaho Peoples, as well as those of the relocated Ho-Chunk, Sac and Fox, and Iowa Peoples.

LIBRARY OF CONGRESS CATALOGING-IN-PUBLICATION DATA
Names: Consolo-Murphy, Sue, 1955– author.
Title: The bears of Grand Teton: a natural and cultural history / Sue Consolo-Murphy.
Description: Lincoln: University of Nebraska Press, [2025] | Series: America's public lands | Includes bibliographical references and index.
Identifiers: LCCN 2024033889
ISBN 9781496236272 (paperback)
ISBN 9781496242914 (epub)
ISBN 9781496242921 (pdf)
Subjects: LCSH: Grizzly bear—Wyoming—Grand Teton National Park—History. | Grizzly bear—Ecology—Wyoming—Grand Teton National Park. | Bear populations—Wyoming—Grand Teton National Park. | Grizzly bear—Wyoming—Grand Teton National Park—Management. | Wildlife conservation—Wyoming—Grand Teton National Park. | Consolo-Murphy, Sue, 1955–
Classification: LCC QL737.C27 C663 2025 | DDC 599.78409787—dc23/eng/20240731
LC record available at https://lccn.loc.gov/2024033889

Designed and set in Garamond Premier Pro by Katrina Noble.

*For my dad, who led our family on early sojourns to parks
and never ceased being excited at seeing another wild animal,
and for my mom, who instilled in me a love of nature and
would have been great editorial help had she lived to read this book.*

Contents

Illustrations

FIGURES

MAPS

TABLES

Acknowledgments

This exploration of history has relied on so many sources and people, all of whom were more than cooperative in answering my questions and sharing their stories. For each of them, I am grateful, and I apologize in advance if I omit anyone who has been helpful. Special thanks to: Paul Schullery for initial and ongoing encouragement, early reads, and reviews; NPS archivist Lynn Marie Mitchell and her team who, along with former park museum curators Alice Hart and Bridgette Guild, gathered and organized the Grand Teton archives and helped me find my way through its priceless records with patience and camaraderie—while still on the park staff and during several years of COVID-19-related closures, Bridgette particularly spent countless hours searching and scanning documents for me when I could not access them in person. I value park colleagues Mallory Smith and Patrick Hattaway for sharing their friendship, their home, and their own memories of bear management; and the overall cooperation I received from National Park Service staff, particularly conversations with and invaluable data from Kate Wilmot and Justin Schwabedissen at the Branch of Fish and Wildlife Management as well as support and reports provided by writer-editor and friend Holly McKinney. Joey Nadeau provided GIS data layers instrumental to making illustrative maps, which were prepared by Megan Smith of Eco-Connect Consulting in Jackson, Wyoming. I am forever grateful to retired wildlife biologist Steve Cain, who patiently ensured accuracy in answering my questions and who, most importantly, oversaw and expanded the park's wildlife program for twenty-three years and kept the records to leave its stories for posterity.

Other former colleagues who shared what amounts to centuries of experience with bears in the ecosystem include Dr. Frank van Manen, leader of the USGS Interagency Grizzly Bear Study Team; Mark Bruscino, Steve Kilpatrick, Dan Thompson, and Mike Boyce, either formerly or, as of this writing, still with the Wyoming Game and Fish Department; Dan Tyers, U.S. Forest Service grizzly bear habitat coordinator; and Dr. Chris Servheen, who for thirty-five years served as the U.S. Fish and Wildlife Service's grizzly bear recovery coordinator. Over four decades I especially benefited from countless lessons about bears through discussions and field time spent with Grizzly Bear Study Team members Mark Haroldson, Craig Whitman, Dan Reinhart (who went on to work at Yellowstone and Grand Teton National Parks), and Jamie Jonkel, who advanced to the Montana Department of Fish, Wildlife, and Parks. I cherish early inspiration and valuable lessons gained about bears and bear management from the late Gary Brown and the late Bob Barbee when I worked for both of them at Yellowstone National Park.

Larry Rockefeller was gracious in sharing personal stories of time on their family's ranch from himself and family members Steven and Mark Rockefeller. Dr. Hank Harlow hosted me for a great interview at his lovely home, as did Jeff Willemain, Wildlife Brigade volunteer and former board member of the Grand Teton National Park Foundation, and retired Yellowstone ranger Bonnie Gafney Whitman, with whom I share memorable bear trapping adventures. Former Grand Teton Park employee Al Williams helped me connect with retirees Barry Alexander, Steve Baldock, and John McAvoy, who recalled otherwise undocumented information on park dumps and trash removal operations. Jackson, Wyoming, police chief Michelle Weber kindly told a personal story of bear #399. Former Yellowstone Park historian Lee Whittlesey graciously shared his history of wildlife and other information. Katherine Wonson, formerly director of the NPS Western Center for Historic Preservation, provided contact information for Dr. Roger Butterbaugh of the White Grass History Project, who connected me with former White Grass Ranch ranch owner Cynthia Galey Peck and Judith Schmitt, Fred Herbel, Karen Gottlieb—workers and guests who gladly shared recollections of their time in the earlier park.

Others who provided helpful information include Chief of Resources Jennifer Carpenter and wildlife biologists Kerry Gunther and retiree Rick Wallen from Yellowstone National Park; Chuck Schneebeck and Kyle Kissock of the Jackson Hole Wildlife Foundation; Jason Wilmot of the Bridger-Teton National Forest; and John Turner, co-owner of the Triangle X Ranch. Leslie Mattson, director of the Grand Teton National Park Foundation, and her staff provided records on the foundation's support for bear science and management. David Diamond, executive coordinator of the Interagency Grizzly Bear Committee, shared access to the records of the IGBC and the Yellowstone Ecosystem Managers' Subcommittee. Mary Centrella; Dr. Susan Clark of Yale University and the Northern Rockies Conservation Cooperative; Dr. Robert (Bob) Smith, long-term colleague and neighbor in Moose; historian and author Dr. Robert (Bob) Righter; former Grand Teton superintendent and acting NPS director David Vela; and writer Todd Wilkinson provided thoughtful insight, encouragement, and tips, as did long-time friend and coworker Gary Pollock, who deserves special mention for generously sharing his photographs for this book. And just when I despaired of finding a cover image representing the subjects of the book, friend and NPS colleague Deb Frauson provided the perfect picture of bears juxtaposed in front of the iconic Teton Range.

This history was also informed by reference materials and images made available to me by historian Brian Beauvais and librarians of the Park County, Wyoming, Library and Archives; Morgan Albertson Jaouen and former employee Nora Haskell at the Jackson Hole Historical Society and Museum; Anna Taylor of the Valley of the Tetons Library in Victor, Idaho; the McCracken Library at The Buffalo Bill Center for the American West in Cody, Wyoming; the National Archives and Records Administration; and the Yellowstone National Park Archives. Two anonymous reviewers, copy editor Stephanie Marshall Ward, and the University of Nebraska Press's editors Kayla Moslander and especially the ever-encouraging Clark Whitethorn helped me make this manuscript better, though omissions or errors of fact are all my own.

And as always, I appreciate the support of my family, near and far.

THE BEARS OF GRAND TETON

I

A Bear Beginning

I picked up my office phone at Grand Teton National Park headquarters one morning in late spring of 2006. On the other end was a rather excited-sounding ranger from the north district, telling me there was a bear at what we called the Colter Bay "C" store, the convenience store and gas station at the junction of the main park road and the road into the major developed area.

"Slow down," I said, "Do you mean a bear has broken into the store?"

No, the ranger said emphatically, but just that morning, several park employees had seen a grizzly bear mother with cubs born that year. For many decades, it had been a rare sighting along one of the park's roadsides. It was the first I could recall hearing about while working in the park.

"Maybe she's just passing through," the ranger said.

I sighed and said, "If it's a bear with cubs, she's not passing through; she lives here," not yet consciously recognizing that the news signaled so much change that had and would come to the valley of Jackson Hole, Wyoming, and the Tetons.

The mama bear was known to wildlife managers and researchers by her number, #399 in the chronological sequence of grizzlies trapped and radio collared by the Interagency Grizzly Bear Study Team, established in 1974 to monitor animals in the greater Yellowstone area. When the team was established, the ecosystem's grizzly bear population was thought to be less than

250 animals and declining, residing mostly within the boundaries of Yellowstone National Park to the north. In 1975 grizzly bears were listed by the U.S. Fish and Wildlife Service as threatened, and for decades afterward, few people reported seeing them in Jackson Hole until grizzly bear #399 turned up and eventually put Grand Teton National Park on the map as a home for her and her kind.

For some thirty-five years, I had the great privilege to work in the national parks of the Greater Yellowstone Ecosystem, millions of acres of mostly wild land in the corner of northwest Wyoming and adjacent areas of Montana and Idaho. In late October 2003 I left northern Yellowstone, where I had been stationed for sixteen years, and moved 145 miles south to Grand Teton Park headquarters. In your car, your map, or your mind, follow the road south through Yellowstone's high forests, broken by steaming geothermal features, through the John D. Rockefeller, Jr. Memorial Parkway, managed by Grand Teton Park's superintendent and staff. Keep heading south along the shore of Jackson Lake, the mighty Teton Range pushing skyward to the west. Then follow the aptly named winding Snake River to Moran Junction—also the southern boundary of the Grizzly Bear Recovery Zone and then the center of bear #399's home range—and on to Moose, a loose concentration of privately owned buildings, a post office, and mostly government houses and workplaces. There, by mid-2006, I settled in as chief of the park's science and resource management programs.

When I arrived at Grand Teton, bear management did not make the list of top priorities, something that would have been unthinkable in Yellowstone, where the animals had commanded considerable public and bureaucratic interest for decades. The acting superintendent handed me a list of four items he expected me to work on in my first year or two. They were, not necessarily in order, to help update the park's fire management plan; to complete environmental compliance for proposed park housing; to develop a new agreement with the long-standing research station in the park, operated by the University of Wyoming; and to tend the often-neglected cultural resources—historic structures, associated cultural landscapes, and a museum collection of Native American art and artifacts in need of conservation and repair. A permanent staff of three biologists and a technician competently monitored newly reestablished wolf packs, studied bison, and addressed concerns over potential brucellosis transmission to livestock in and adjacent to the park. We were also working on, with the neighboring National Elk Refuge, a long-term bison and elk management plan—a seven-year effort finally completed in 2007. And they dealt with bears when necessary.

The park was also fleshing out the transfer of the last piece of the privately owned JY Ranch to the U.S. government, in keeping with the wishes of longtime owners John D. Rockefeller Jr. and his son Laurance. Staff were assessing options for transit and improvement of park corridors such as the Moose-Wilson Road. Managers struggled with how to convert an ill-planned addition to the park's maintenance building into a badly needed new headquarters and how to fund this conversion. They also struggled with how to consolidate staff spread out in various temporary and unsustainable structures. Under tightening budget scenarios, the regional director had directed parks to assess their "core operations," a painful administrative exercise that evoked emotions from staff in all disciplines as to their relative worth and their fiscal needs compared to their colleagues. No one lacked for tasks to keep them busy.

Bears were not high on the list of things to be concerned about in the early twenty-first century Tetons. Compared to its more famous neighbor, Grand Teton National Park was known as a safer place to hike, camp, and backpack, largely because one was unlikely to encounter or even see a grizzly bear. Indeed, until 2006, I had only one fleeting glimpse of a grizzly in the park and recall no incidents with them. Grand Teton and the Rockefeller Parkway had bear management plans and programs to make trash unavailable to animals, similar to efforts at Yellowstone. Black bears were commonly seen and occasionally needed to be deterred or removed from a campground or a busy trail. Visitors were understandably frightened and rangers responded, but typically without controversy.

Much has been written about "Yellowstone" bears. Who doesn't know that, from early in the twentieth century, visitors viewed bears—mostly black bears, one of two species found in America's lower forty-eight states—

from park roadsides and in garbage dumps, or at least that the cartoon Yogi Bear faced (and generally bested) his ranger nemesis in Jellystone Park? The region's other bear is the grizzly, and the Yellowstone area is home to the most studied population of large carnivores anywhere in the world. Grizzly bears once ranged from the Pacific coast of Canada and the United States to the Great Plains and southward into Mexico. But their distribution and numbers, like bison and other once-more-widespread large mammals, shrank as pioneers moved across the plains clearing forests, establishing farms and ranchlands, and defending themselves and property with the most lethal enemy a grizzly bear has—firearms. By the 1920s grizzly bears, also known and classified by biologists as brown bears, were gone from the west coast of the coterminous United States (the states outside of Alaska and Hawaii) and functionally from the Southwest and Mexico.[1] For a time, grizzly bears persisted in the Pacific Northwest, western Montana, and remote mountains south to the Yellowstone ecosystem, including the Teton Range.[2] However, by the 1940s, in the wake of human settlement, the larger of the two continental bear species remained only in and around Glacier and Yellowstone National Parks; though black bears were consistently present, the grizzlies were essentially gone from Grand Teton National Park.

Like me, other Americans formulated their conservation interests by watching, among other things on television, twin brothers Frank and John Craighead capturing and tracking those first wild brown bears in the heart of Yellowstone Park. Many people followed the 1970s drama surrounding the end of their research, as unrecognized government bureaucrats ceased support for the study. Was it because the scientists disagreed with the National Park Service's larger estimate of how many bears remained, and with the decision to abruptly remove the garbage dumps around which the study had been centered? Was it even because one of the brothers testified against the Park Service in a court case about the killing of a park visitor by a grizzly in 1972?[3] And what would become of the grizzly bears, subsequently listed for protection under the Endangered Species Act—could they survive without the robust supply of park trash that had helped feed them for decades? Yellowstone Park has not lacked for controversy, even drama, surrounding its famed bears.

In contrast, Jackson Hole—the broad, glaciated valley cut through by the Snake River east of the Teton Range, which rises abruptly, without foothills, from the sagebrush plain—is best known for spectacular scenery, for long winters, and for recreation such as horseback riding, mountain climbing, fishing, boating on Jackson Lake, river rafting, and skiing big alpine terrain. Nineteenth-century trappers of beaver and others named the "hole." But for much of its history, if Grand Teton National Park was known for wildlife, it was mostly for elk, which were popular with hunters and with visitors, who could see thousands of them wintering on the National Elk Refuge bordering the park.

Then, in the late 1900s, the area began to experience major changes in how both humans and wildlife used and appreciated the Teton landscape. And as the twenty-first century progressed, grizzly bears became renowned in Grand Teton National Park, enough to prompt questions that I, while working, seldom had time to explore: Why were people—from the ranger who called in May 2006 to other local residents and park visitors—surprised to think of grizzlies in Jackson Hole? How many bear "problems" had occurred over time, and how had the park handled them? Where had the park historically dumped its garbage, and had bears come to those garbage dumps and panhandled along park roads as had happened so famously in adjacent Yellowstone National Park? How had livestock grazing, which was still permitted in the park, been affected by bears, if at all? What about the seldom acknowledged Rockefeller Parkway, which connects the two parks, and where hunting remains legal to this day—how did or would that affect bears in the past, present, and future?

As far as I could find, the history of bears in Jackson Hole had not been compiled in one writing. This, then, is my intended focus, to add to the story of the ecosystem with a history of bears and of humans' responses to them in what has become Grand Teton National Park and the John D. Rockefeller, Jr. Memorial Parkway, with a little spillover into surrounding areas. The core setting is roughly bounded on the west by the Teton Mountains straddling the Wyoming/Idaho border, extending north to the southern boundary of Yellowstone, then following the Snake River valley's eastern edge to the

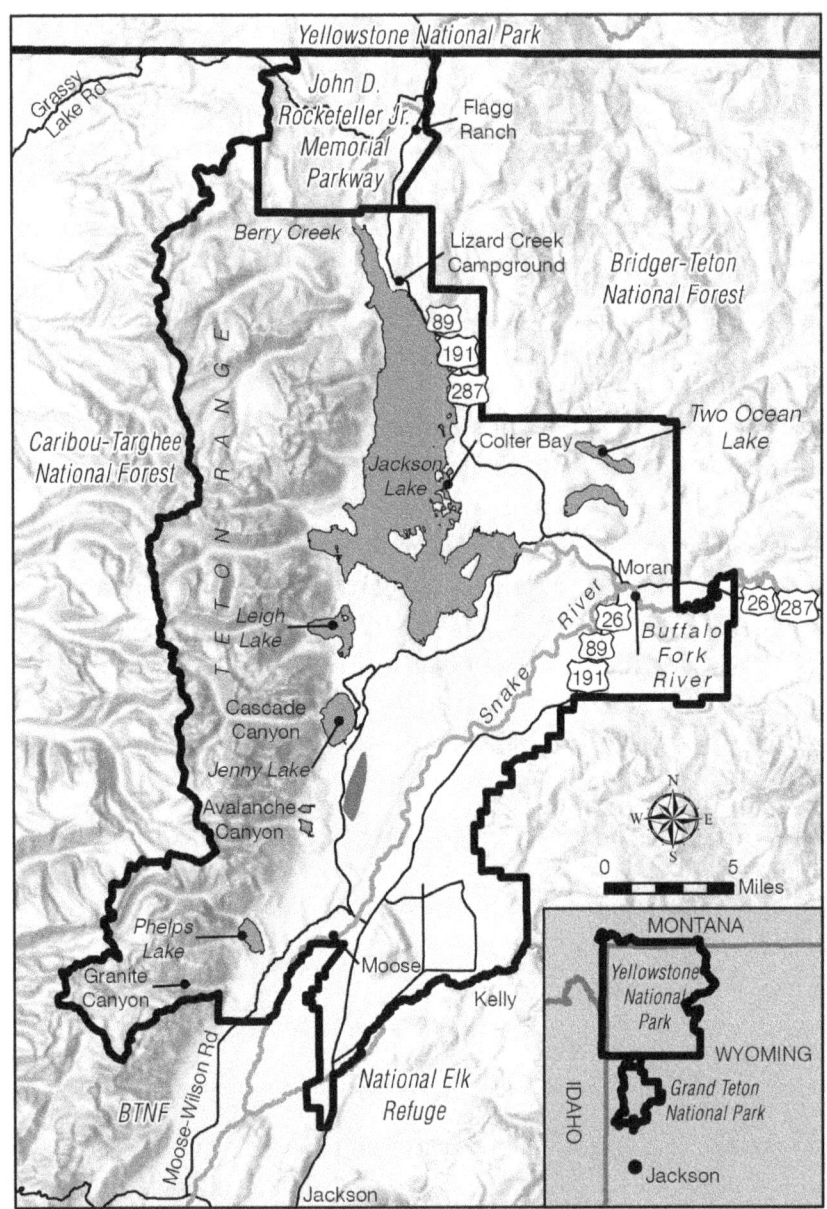

MAP 1. Orientation, Grand Teton National Park, John D. Rockefeller, Jr. Memorial Parkway, and surrounding area. Map by Megan A. Smith, EcoConnect Consulting, Jackson WY.

valley of the lower Buffalo Fork River, on to the base of the Gros Ventre Range, and south to the town of Jackson.

My career allowed me twenty years with varying responsibilities on behalf of public education and bear conservation and management in what some, in and out of the National Park Service (NPS), refer to as "the mother park." I was used to the verbal and literal arm waving that assures audiences that what is said about "Yellowstone bears" encompasses what has become known as "*greater* Yellowstone," including Jackson Hole. The species' biology is essentially the same across the ecosystem, so my intent in this work is not to repeat much of the ecological information that is well told in general works about bears, such as the recent *Yellowstone Grizzly Bears*.[4] But beyond biology, Yellowstone National Park differs from areas outside its borders. It is different in its age, having been set aside by the U.S. Congress in 1872, from land unoccupied by Euro-Americans, and not even within one of America's united states, as the world's first national park. Its lands have essentially never been settled or developed for year-round occupation, save by a small number of park employees concentrated in a few areas. The park is thought from its inception to have retained—with the notable exception of wolves, which were restored in 1995–96—all the native species, including bears, that had existed when Anglo-American explorers first "discovered" the wonderland of geysers and hot springs. The NPS is the U.S. government entity delegated almost sole responsibility for over 2.2 million acres of public land therein. Yellowstone National Park has led the world in setting up bear management and research initiatives that provide models for other places. Yet it is, dare I say, blessedly lacking in some managerial complexities facing other national parks as well as national forests that have private or state lands, residences, and those associated jurisdictions, laws, and administrators within their boundaries.

The history of bears and bear management in Grand Teton National Park and the parkway is emblematic of areas beyond Yellowstone, where the bruins existed before but declined or disappeared in the face of human settlement and land use, hunting, trapping, farming, and ranching. It and other bear-occupied national parks, like Yosemite and Glacier, do have considerable legal protection and mandates to conserve wildlife and other resources, but by the time these other parks were established there were private residents and businesses working the landscape as well, influenced by and influencing the native animals. In Grand Teton, the Rockefeller Parkway, and elsewhere in bear country, there are countless areas in which guest ranches, private homes, and hunting and livestock grazing or other agricultural uses tempt bears into conflicts with humans, conflicts to which communities and users and managers of state and federal lands have responded, in the past and present.

Over my decade-and-a-half tenure in Grand Teton National Park, bear management became the priority issue it had not been before. In this place, one highly visible and publicized bear, #399, billed as "the world's most famous grizzly bear," has drawn new attention to Grand Teton National Park, to her species' status, and to the challenges of such large, dangerous animals coexisting with increasing numbers of people living in and visiting wildlands.[5] While bear #399 is not the focus of this book, neither is she absent from it, since her presence on the landscape commanded considerable attention by the public and staff, in the park and beyond. I encourage readers to consider how she or any other individual bear represents and differs from others of her kind.

This book is a combination of personal recollections, mostly but not solely my own, and research from archival records, old newspapers, and other publications and writings. I view it as a work in progress and am open to additions and revisions to things that are incomplete or wrong. Since history continues to be made every day, a written one becomes dated as soon as it is published. The story of bears in Jackson Hole is ever-changing and surely will grow with even more rich tales in the future, beyond tales of any single bear or bear manager.

2

Where Bears Belong

More than once in my years at Grand Teton National Park I recall being summoned to talk about bears, along with senior wildlife biologist Steve Cain, in the office of the park's first female superintendent, Mary Gibson Scott. Like any good manager, Mary, as she settled into her position, made a point of getting to know the community and its leaders and influencers. Many of them, in turn, did not hesitate to reach out to her and share their opinions about the park and how it should operate. Although I don't recall this ever happening with black bears, after 2007 or 2008, we would hear from locals as grizzly bears occasionally moved south outside the unfenced park boundaries, following the riparian zone of the Snake River. It's fine wildlife habitat, with some large tony homes on spacious residential lots among the Douglas firs and cottonwoods lining the wide floodplain. With her booming laugh, which I assume she did not share with callers, Mary would tell us when a park neighbor phoned to report that "one of our grizzly bears was out of the park and on their property, and would we come and take it back where it belonged?" (The park would, politely, "just say no.")

Both black and grizzly bears lived in Jackson Hole long before its settlement. Beginning at least 11,000 years ago, post-glacial people traveled over mountain passes in the Teton Range to camp, fish, collect plants, and hunt along the Snake River and on the now inundated shores of the natural Jackson Lake.

American Indian tribes traditionally associated with the area include the Shoshone, Blackfeet, Crow, and numerous others who provide some indication of early awareness and use of bears. American writer Margaret Sanborn, referring to the Shoshoni band of Sheep Eaters as Tukadüka ("Tukudika") noted that in what's now greater Yellowstone, tribal hunters caught bears in pitfall traps, little more than holes dug in the ground covered by branches. Throughout the Rocky Mountain West, Indians used bear fat as a medicinal salve and strung the animals' claws into necklaces. Bears were among the frequent guardian spirits for Indians undergoing vision quests.[1] David Rockwell, in *Giving Voice to Bear*, wrote that a bear-knife bundle, containing a long-handled blade made from a bear's jawbone, was a symbol of power to the Blackfeet and their allied Gros Ventre tribe, and that Crow chief Plenty Coups ate grizzly bear meat as a youth to become a strong warrior.[2] Although ethnographic studies of Grand Teton National Park are not public documents, there are assuredly many other rich stories Native Americans could tell about their long and continued relationship with bears in the area.

The earliest Euro-American references to bears in or near Jackson Hole appear to be from trappers probing then-unsettled river valleys for beaver and other furbearers, such as the Wilson Price Hunt party sent by John Jacob Astor to explore west from St. Louis and establish a fur trading post on the Columbia River. On September 28, 1811, Hunt and a companion moved twelve miles upstream on the Snake River from near present-day Hoback Junction to the southern end of Jackson Hole, where they reported "many indications of beaver . . . a band of elk, and . . . two bears, one black and one gray."[3] Hunt might have been referring to different colors or species of bears. Early references to bears in the western United States, from the journals of Lewis and Clark on, not uncommonly mention the "white" bear as synonymous with "grizzly," but it is less conclusive that a "gray" bear means the same.[4] It is also common that early writings refer to "bear" versus "grizzly," suggesting that, as often as not, a simple "bear" reference meant a black bear. Either way, Hunt had multiple sightings of one or both native ursids.

Along southeastern Jackson Hole runs the Gros Ventre River. Its wide, mostly open valley of sagebrush grassland is winter range for thousands of elk and deer, and the Jackson pronghorn herd migrates through it in summer and fall. Warren Angus Ferris, surveyor and trapper with the American Fur Company, who later briefly explored Yellowstone, camped in the Gros Ventre valley on May 31, 1833, and killed a large bear of unnamed species.[5] A compatriot of a young trapper named Zenas Leonard had a fatal close encounter with a grizzly the following year, after traveling several days north from Star Valley, Wyoming:

> On August 8, near the Snake River, "passing through a small prairie, we discovered a large grizzly bear laying in the shade of some brush at the edge of the woods, when four of us started for the purpose of killing him, but on coming close, the bear heard us and ran into the thicket. We now took separate courses, intending to surround the bear and chase him out and have some sport; but one man, as we came to the thicket, acted very imprudently by dismounting and following a buffaloe path into the brush, when the bear, hearing our horses on the opposite side, started out on the same path and met the man, when he attempted to avoid it by climbing a small tree, but being too closely pressed was unable to get out of reach of the bear, an as it passed, caught him by the leg and tore the tendon of his right thigh in a most shocking manner. Before we could get to his aid the bear made off and finally escaped." The man died the next day, and after his burial his party moved on and "encamped on the headwaters of the Lewis River [in southern Yellowstone]."[6]

Historical tales exist of bear encounters with trappers and early explorers in both valley and mountain areas of northwestern Wyoming and the Idaho side of the Tetons. "Beaver Dick" Leigh, for whom Leigh Lake in the park is named, recorded numerous bear sightings in the Teton Basin along the streams west of the range and wrote that in "the butifull valley of Jackson Hole . . . the game . . . is not to be snesed [sneezed] at for elk and deer and Bare is very plenty."[7] The first evident reference to a bear sighting in today's Grand Teton National Park comes from Nathaniel P. Langford's expedition to the region in 1872, the same year Congress designated Yellowstone Park based largely on witness and photographic evidence from Langford's and others' explorations. Whittlesey and Bone, in their thoroughly researched history of mammal sightings in the region, report on the excursion of the southern division of the Hayden Survey that they approached the Tetons from the west, some of them thinking of climbing the Grand Teton, and:

> On July 24 . . . After the party left their horses and climbed to an elevation of about 10,500 feet, Langford said they could see "a large lake of marvelous beauty . . ." They named this Cowan Lake . . . now known as Snowdrift Lake (Bonney and Bonney 1992). [8] . . . "in that portion of the Teton Range known among the early trappers as Jackson's Hole." They chose a spot to camp that night, and "When within three miles of it, we came upon our fearless topographer, Mr. Beckler [*sic*: Bechler], who, with a shotgun loaded with small shot, stood face to face with a she grizzly and two cubs, which he had frightened from their lair in the thicket . . . Fortunately, in attempting to discharge his gun it missed fire, and probably saved him from a deadly encounter with the irritated animal, or a hasty ascent of a tree as a possible alternative. We prevailed upon him to return with us, and await a more favorable opportunity for a tussle with grizzlies. In camp that night they learned that 'Two other members of our party had killed and brought into camp a good-sized black bear, which is one of the most formidable animals in the Rocky Mountains.'"[9]

Snowdrift Lake lies high in the north fork of Avalanche Canyon, a major drainage on the east side of the Teton Range. Even today it is accessible only on unmarked hiking routes and to hardy people, such as those willing to bushwhack through a boulder field in the southwestern part of Grand Teton Park. It's unclear whether this is the same lake where party member Sidford Hamp saw tracks of a large grizzly bear several days later.[10]

During that expedition, renowned photographer William Henry Jackson captured his recollections of wildlife while taking the first images of the Tetons:

It was a perfect day, clear and cold, but with enough warmth in the sun's rays to melt the snow in trickling rivulets on the southerly exposures, thus keeping up the water supply required for plate washing . . . it happened that I was alone for some time, wrapped up in my tent preparing a plate for the next view . . . I was astonished to see a magnificent mountain sheep—a buck with enormous horns—on the rocks within thirty feet, watching my movements with apparent fascination . . . His presence there was evidence of the extreme isolation of this region, for it was, perhaps, one of the least frequented places in the Rocky Mountains. Sheep and bears were numerous all over the higher plateau, and elk, deer, and moose lived in the wooded seclusion of the canyons and hillsides . . . The Teton region at this time was a game paradise. Our various parties were kept supplied with fresh meat without having to hunt for it . . . It was equally easy to get a mess of trout from the streams nearby. Bears were abundant also. The first day in the main camp, two of the younger boys were fishing, and unexpectedly happened on bruin. This was larger game than they expected to meet, but they succeeded in killing the bear with pistols only.[11]

The survey party's comments on the abundance of bears and other wildlife in the Teton Range typify what most early visitors reported. Joseph Mushbach, assistant topographer for the 1878 Hayden Survey, camped in August and journaled, "On the Snake, just above the mouth of Great Bend Creek . . . Judging from the sign yesterday, bear must be as thick as rabbits as [sic: at] home, in the Teton bottom."[12] Great Bend Creek is a tributary to the Snake River not referenced elsewhere, but Mushbach's group had camped the previous night "about on a line between the North end of Upper Big Gros Ventre Buttes, and the South Teton" . . . and the following night "on the Snake about 5 miles above the head of Jacksons Lake," placing this tale clearly in the valley, likely within the current boundaries of Grand Teton National Park.

William Baillie-Grohman visited the Tetons during a late-nineteenth-century journey through the Rocky Mountains and, while exploring the Tetons, liked to fish with "hopper-bugs" particularly "at the outlet of Jennie's Lake . . . for what he called salmon [cutthroat] and eat them fried in bear fat . . . a right royal dish." While fishing one day near dense serviceberries, he had what he called a

> ludicrous little incident . . . when out of the bushes, as if growing from the earth, there rose—a grizzly. Rearing up on his hind legs, as they invariably do on being surprised, he stood, his head and half-opened jaws a foot and a half or two feet over my six foot of humanity, and hardly more than a yard between gigantic him and pigmy me . . . he looked the biggest grizzly I ever saw, or want to see, so close. It would be difficult to say who was the more astonished of the two, but I know very well who was the most frightened . . . Now grizzly shooting is a fine healthy sport—I know none I am fonder of; but . . . I was here on a treeless barren . . . and yonder, 100 yards off, lay [my] famous old rifle . . . Fortunately . . . I saw very plainly that he was more puzzled as to my identity than I was regarding his. His small, pig eyes were not very ferocious-looking, and first one, then the other, ear would move; expressing, as I interpreted it, more impatience than ill-feeling. I do not exactly remember who first moved, but I do recollect that on looking back "over my shoulder" I saw the old gentleman actually running away from me! . . . This is not the only bear story I could tell, but . . . I will not weary the reader's patience with what has been told so often, namely that grizzlies want no fooling.[13]

After the nineteenth-century forays of westward explorers, trappers, and government survey parties, the less hospitable landscape at Yellowstone's higher elevations was and would remain unavailable for settlement. But in Jackson Hole, humans kept coming, and settlers of the first permanent homesteads, established in 1884, tried to carve out a living farming and ranching up and down the Snake River and its tributaries.[14] In or after 1896, Francis Judge, referring to a location in Buffalo Valley, mentions the road down Uhl Hill (near Moran Junction and the Elk Ranch in today's park) to her "Gram and Gramp's ranch," where "every year or two a bear would be killed. The grease—pale yellow and soft like honey—would be used, among other things, for deep frying. Gramp was sure that no fat could equal bear grease for doughnuts. And he always carefully skinned and prepared the feet for pickling. What a rare treat—pickled bear's feet!"[15]

At the turn of the nineteenth century, more than six hundred people lived in the valley. Living was not easy, and residents' use of and interest in bears varied from the practical to the curious. Owen Wister, author of *The Virginian*, spent several months with his family in 1911 visiting the valley's first dude ranch, the JY, located off the Moose-Wilson Road in what is now Grand Teton National Park. By 2007, buildings at the JY were removed to establish what is now called the Laurance S. Rockefeller Preserve. The rich wildlife habitat includes glacial Phelps Lake and mature mixed-conifer forests extending the short distance to the Snake River, which comes as close to the base of the Teton Range as anywhere along its forty-mile length. It is fair to assume that the ranch, when owned by the Rockefeller family, offered high-quality experiences for its private guests. But Wister's family had a quite different memory of the fare, which was likely hunted as close as possible to the ranch: "Wister's daughter, Fanny Kemble, recalled that food at the JY was often scant, and not always the best. On those days 'when a steer was shot for beef, we would have some of it for supper . . . We ate dried, smoked, salted bear [like dark brown leather] from the year before; [and] fresh elk too tough to chew . . . We frequently found dead flies between the flapjacks at breakfast, and we drank condensed milk.'"[16]

Valley residents formed several small villages, only some of which remain. Jackson, named in 1894 and destined to become the largest, was "still in essence a frontier community" with a post office, Deloney's mercantile store, and a recreational hall located near today's town square.[17] Photographs of the town show that by 1909 a loose amalgam of several dozen buildings stood midway between Snow King Mountain and East Gros Ventre Butte. Yet five years before the town was incorporated, locals could enjoy news in the *Jackson's Hole Courier*. In its first decade, it regularly printed briefs such as these about hunting for sustenance or sport:

Wm. Crawford returned Tuesday, from a several days' bear hunt, which, however, failed to prove very successful.[18]

George Erwin, who recently filed on land on the Buffalo [Fork], went into the timber the day before Christmas to cut some poles and work up a Christmas appetite, and while there ran across a large bear enjoying the sunlight. The result is that George has about eighty pounds of "bar grease" and a nice six-foot rug to land on in the morning.[19]

Charlie Sowers, fourteen year old son of Mr. and Mrs. S. S. Sowers, demonstrated a couple of weeks ago that age isn't a necessary requisit of a good hunter. He had gone back on Sheep Creek with his two dogs and his rifle, and there came across a couple of old bear and two cubs. Charlie got the mother bear the first shot, but the other old bear got away. The two cubs climbed a tree and, as he didn't like to kill them he tied his dogs to the tree and returned to the ranch for assistance. His father accompanied him to the spot the next morning and after much difficulty they got the cubs into a sack and took them home. The hide of the old bear measured more than seven feet in length and about six and a half across the shoulders, a pretty fair bag for a fourteen year old youngster.[20]

Bruce Porter, the up-to-date young druggist, has returned from a hunting trip, bringing down a moose and a bear. Several days ago, there was a report about town that Ole Warner, the intrepid bear hunter, was chased up a tree by a bear and that Bruce was fortunately, near at hand and made a thrilling rescue. We ourselves, knowing Ole as we do, doubt the truth of such a report. (However, the following week's issue of the Courier had this postscript: Anyone wishing to see the bear that treed Ole Warner just drop in at the Drug Store.)[21]

FIG. 1. Undated photo of Jackson mayor Grace Miller with her pet black bear. Collection of the Jackson Hole Historical Society and Museum, 1958.2881.001.

The relationship between early Jackson Hole residents and bears was not exclusively about hunting. Several locals had pet black bears, which does not appear to be an uncommon practice. One of Yellowstone's first superintendents, Captain George Anderson, kept a black bear chained to a pole outside his Fort Yellowstone quarters in 1890. Early Wyoming was more egalitarian than most states and, in 1920, Jackson voters elected the first all-women's ticket in the United States for town council and mayor. The new mayor, Grace Miller, had a pet bear.[22]

T. N. McCoy of Wilson, Wyoming, placed an ad in the paper, "BEAR FOR SALE—One year-old, pet, black bear. Will sell him very reasonable."[23] In 1926 Harry and Ethel Harrison of the Circle H Ranch, on the Snake River just south of present-day park headquarters in Moose, had a pet black bear named Ming who lived under their front porch.[24] And one of the valley's earliest homesteaders, John Cherry, told this possibly true tale of a

grizzly bear that he had raised from a cub. He kept him on a chain at his cabin when he was little . . . and made quite a pet of him. A mature boar grizzly will weigh 900 to 1000 pounds, the weight of a middle-sized horse; so when the cub grew up John "taught him to ride," and they'd really sail around the forest. Well, John liked to hunt and the bear liked to hunt, too; so they made quite a pair. John'd grab his rifle, jump on his bear, and off they'd go. One day when they were out hunting, they came upon quite a group of grizzlies feeding on grubs and berries. Well, John didn't want his bear getting mixed up with that bunch; so he jumped off and emptied his rifle into the air to scare the others off. For a few minutes there was quite a commotion, with bears going every which way. When they had all left, John jumped back on his bear and started for home; but they hadn't gone very far when he realized he was on the wrong bear.[25]

Early newspapers also provided stories of wildlife around the world, of natural history (whether accurate or not, based on current knowledge), and of early thoughts of conservation—a reminder that people a century ago, just as today, had wide-ranging viewpoints on bears and their relationship to human society.

GRIZZLY BEAR FROM ASIA—Something Like a Million Years ago the Animal Came Here by Way of Alaska: The grizzly bear has been known to the white race little more than a century. Lewis and Clark wrote the first official accounts of him in 1805 . . . Guthrie's old geography says that he was named Ursus horribilus by Naturalist George Ord in 1815 . . . He appears to have come into America about a million years ago over one of the prehistoric land bridges that united Alaska and Asia. Bears and dogs are descendants from the same parent stock. The grizzly bear never eats human flesh, is not ferocious and fights only in self-defense. He leads an adventurous life, is a born explorer and ever has good wilderness manners—never makes attacks. The numerous cases in which the grizzly has been made a pet and companion of man, where he was thoughtfully, intelligently raised, show him to be a superior animal, dignified, intelligent, loyal, and uniformly good tempered. Not a grizzly exists in any of the four national parks of California, and that animal, once so celebrated in that state, is extinct there. He is also extinct over the greater portion of the vast territory which he formerly occupied, and is verging on extermination.[26]

There was still an air of fear and often clear dislike, based on the animals' ability to compete with residents for their livelihood. Cattle ranching had begun in Jackson Hole in the late 1880s, and it had not yet been supplanted by dude ranching, as it would be in the early 1920s. Neal S. Blair's *The History of Wildlife Management in Wyoming* describes the state's actions decade by decade and, at that time, as the Wyoming Game and Fish Department had yet to be officially established, the game commission was responsible for policy and staffing efforts at wildlife management. At a July 1921 gathering, members of the Wind River Livestock Growers' Association discussed predators, especially bears, and

> whether the legislation can not be altered so as to aid the stockgrower in ridding the country of the black, brown, and silvertip varieties. At present it seems to be the aim of the State [Game] Department to protect these animals. The association has instructed its secretary to send a copy of its resolution condemning the extension of the park; condemning the State Game Department's attitude on the predatory animal regulations and opposing any restrictions to livestock on game preserves . . . a petition and a resolution were drawn up requesting that Harve Burlingham be appointed a state predatory animal trapper.[27]

Despite the reported views of stock growers at the Jackson meeting, one of the earliest references to the state game commission partly counters the anti-predator viewpoint:

> The commission favored strong predator control measures since predatory animals were regarded as a serious menace to herd increases among the game animals. Blame was often hard to place, however, and some

veteran wildlife men such as Jess Stull of Jackson feel the bear has often taken the blame for depredations he did not actually commit. Bears are not known for the delicacy of their appetite and some observers feel they often feed on the carcasses of stock killed by other causes. The bolder and more discriminating will raid camps and food caches as many a sportsman or stockman has learned to his sorrow.[28]

In the early years of the twentieth century, bears in Yellowstone National Park were already becoming a tourist attraction and had to contend with few human residents in their habitat. They were not viewed unfavorably as competing with so-called good animals like deer and elk and were thus not subjected to predator control programs as were mountain lions and wolves. To the south, public sentiment on bears appeared to be mixed; among the 1,400-plus citizens of Jackson Hole were those who saw grizzly and black bears as predators, pests, pets, even something worth protecting, as a 1920s movement toward conservation prompted the establishment of Grand Teton National Park.

3

Bear Management in the Early Park

1929–1950

Early in my years at Grand Teton National Park, the National Park Service (NPS) had an initiative to improve archives in or about the parks. Few of them keep their own natural and cultural resource records on-site as a useful service to researchers, though it's permitted by the National Archives, which oversees and houses most U.S. government record collections. In 2005 we hired a new curator to pack up and transfer a unique American Indian art and artifact collection to a facility in Tucson for conservation treatment. Once that was done, amid the office moves necessitated by vacating old buildings in need of rehabilitation, we began planning to compile the park's own archival records. Well did I recall the Yellowstone tale of engineer-turned-historian Aubrey Haines, half a century before, scrounging files from dusty closets and literally saving boxes of papers from the trash dump to build that park's own archives.

To begin, we announced our interest in old files and offered employees training in how to recognize and save documents important to park history. Some staff were notoriously reluctant to part with "their" files, even if boxes full of papers appeared to have gone untouched for years, gathering dust. The hunt for records sometimes resembled Aubrey's tales. One morning the curator arrived at work to find a file box, carefully wrapped but unidentified as to who had mysteriously left it overnight on her desk, as though she or he had ferreted away precious gems to be secretly sold at a pawn shop. With years of help from a veritable whirlwind of energy in the petite person of regional archivist Lynn Marie Mitchell (since retired) and her able team of assistants, Grand Teton Park gathered, as archivists measure, some 528 linear feet of original materials—nearly 845,000 papers, photos, and other items ranging from mountaineering logs of Teton climbers to wildlife and other resource records, including those related to bear sightings and management. In July 2015 the park proudly christened the new archives, and its contents have served to greatly inform this book, which I did not imagine writing at that time. In poring through the depths of the archives, I marveled at the gems found within, and at the gaps—after a lovely set of forms documenting bear sightings and handlings from 1940 to 1945, did no one keep similar records for the next decade, were they lost, or did nothing happen worth recording?

Land conservation efforts in Jackson Hole predated Grand Teton National Park. In 1897, twenty-five years after the establishment of Yellowstone National Park, President Grover Cleveland set aside the 829,000-plus-acre Teton Forest Reserve. As with the first park, there were few funds or staff allocated to manage the area, but gradually the federal government began to tend their public lands. In 1908, under the young U.S. Forest Service, legislative protection was expanded to the newly named Teton National Forest, nearly two million acres—to be managed for water projects, livestock grazing, timber production, and recreation—encompassing the non-private land in the valley or "hole" and its surrounding mountains. Lively debates (some continuing to this day) occurred over whether public forests should be managed for an array of multiple, often consumptive, uses or whether they should be focused on recreation and "preservation," such as in national parks. Some of the Teton Forest, indeed, later came to be part of the National Park System.

Since Anglo-American settlement began in the 1880s, early consternation over large numbers of elk dying in the harsh winter prompted locals to feed the animals and support the creation, in 1912, of the National Elk Refuge on the northern edge of the town of Jackson. Other wildlife did and do occupy the elevational span of the Teton Range, but throughout the twentieth century elk were the dominant species of interest, especially in the high mountain valley that provides the Jackson herd both summer and, primarily just outside the park, crucial winter range. Bears are seldom mentioned, either as attractant or nuisance, in early local histories; residents were busy building lives and livelihoods.[1]

The establishment and later evolution of the park into what you find on maps and on the landscape today is itself a complex and fascinating story told elsewhere in fine detail by others, especially historian Robert W. Righter in his notable 1982 *Crucible for Conservation*. A much-condensed version is that protection of the scenic valley from burgeoning summer homes and commercial development, and of elk habitat and migration routes southward from their summer range in Yellowstone, prompted early discussion of expanding the world's first national park or creating another. In the 1920s other concerns arose related to damming and enlarging Jackson Lake, and to additional proposed dams at Jenny, Two Ocean, or Emma Matilda Lakes to irrigate grazing lands in Jackson Hole. In February 1929 Congress protected 96,000 acres of mountain land, west of but not including Jackson, Jenny, and other glacial lakes or the Snake River bottom lands of the valley (see map 1), as Grand Teton National Park.[2]

Though it was smaller than what he had hoped for, NPS director Horace M. Albright, in his annual Report to Congress, described the park's establishment and dedication as one of the year's highlights. The first superintendent did not report for active duty until June, due to snows in the high country. In addition to the head man, the park then employed only a temporary clerk and one permanent and two temporary park rangers. It is unclear who on the small staff would have provided information for the initial report that "black bears are abundant in the higher country and grizzly bears are occasionally seen in the deep canyons where the tourists seldom travel. With two or three years of protection they will be as plentiful here as in the Yellowstone National Park."[3] One wonders what prompted this optimistic statement, which unfortunately was not borne out for the better part of another century.

Inevitably the new park added personnel, who provided public education programs and news reports, such as this summer 1937 tale from Grand Teton's *Nature Notes*:

> Two C.C.C. [Civilian Conservation Corps] spike camps have been visited by bears this month. A small black bear carried off several pounds of meat from the Granite Camp. This particular bear showed up at a later date and was chased up a tree by a group of enrollees. One of the boys, bolder than the rest, followed the bear up the tree with a cross-cut saw. The bear perched out on a limb as far as he could go. The enrollee sawed off the limb letting Mr. bear take a good fall. He scampered off through the woods and has not molested the food supply of this camp since.[4]

The casual, humorous tone with which park workers described the bear encounters suggests ignorance of the risk. By this time Yellowstone Park was actively feeding grizzlies at garbage dumps, but it was also experiencing bear-caused human injuries, mostly from visitors feeding bears along park roads and in campgrounds. Managers there received an increase in public letters sent by citizens and elected officials and had begun talking about how to address the associated dangers. That year alone, bears in Yellowstone injured a record 115 people.[5]

Grand Teton's records of bear activity and associated management actions (see table 1, chapter 4) do not start until the park was a decade old. Then for five years some diligent employee(s) recorded animal sightings and conflicts; they immediately suggest some challenges for the young park. In July 1940 rangers trapped a black bear in String Lake campground and released it outside the park south of Wilson, Wyoming. That October bears reportedly caused serious damage to snowshoe cabins in the northern portion of the park.[6] The superintendent's Annual Wildlife Report referenced considerable bear trouble in campgrounds and other points of concentration as well as at neighboring ranches, where two troublesome bears were killed during the summer. The tantalizing note that "there is abundant natural forage and it is hoped that the use of the park incinerator next year will help curb this increasing bear problem" prompts unanswered questions about where and how the early park began dealing with the day-to-day management of garbage and bears.[7] No records, then or later, indicate the existence or acquisition of a trash incinerator. Rangers continued to report increases in bear depredations on garbage cans and ice boxes, especially at dude ranches, but did not mention any actions taken in response. Clearly, Grand Teton was experiencing similar bear-human conflicts to those occurring at Yellowstone, but it appears they did not attract much attention, as the park was occupied by other concerns.

Establishment of the original 1929 Grand Teton Park was followed by a two-decade effort to expand or even add it to Yellowstone. Persistent unease over development prompted initially quiet efforts by John D. Rockefeller, Jr. and partners' Snake River Land Company, actively supported by superintendents, to purchase valley lands that would be donated to the U.S. government and included in an enlarged park. By the 1930s Jackson Hole was occupied by auto camps, homesteads, summer cabins, and ranches, some of which grazed cattle and most of which had by then morphed into sites that more profitably served guest "dudes." It is unclear whether park leaders considered what wildlife management challenges they would inherit on lands that had been privately owned, although long-standing concerns over elk migration and hunting continued among the various jurisdictions. There was considerable drama over national park expansion and plenty of viewpoints ranging from wholehearted support for to armed opposition to an enlarged park. The perceived secrecy with which land had been acquired, and suspicion that some landowners had sold land for less than what they might have received (had they known one of the nation's richest men was making the purchases), added to the vehemence with which opponents, including movie stars and congressmen, fought park expansion.

By the early 1940s, Rockefeller tired of paying taxes on lands he had bought and threatened to dispose of the properties elsewhere if the U.S. Congress did not accept his donation. Since legislators had not acted, Franklin Roosevelt established a 221,610-acre Jackson Hole National Monument

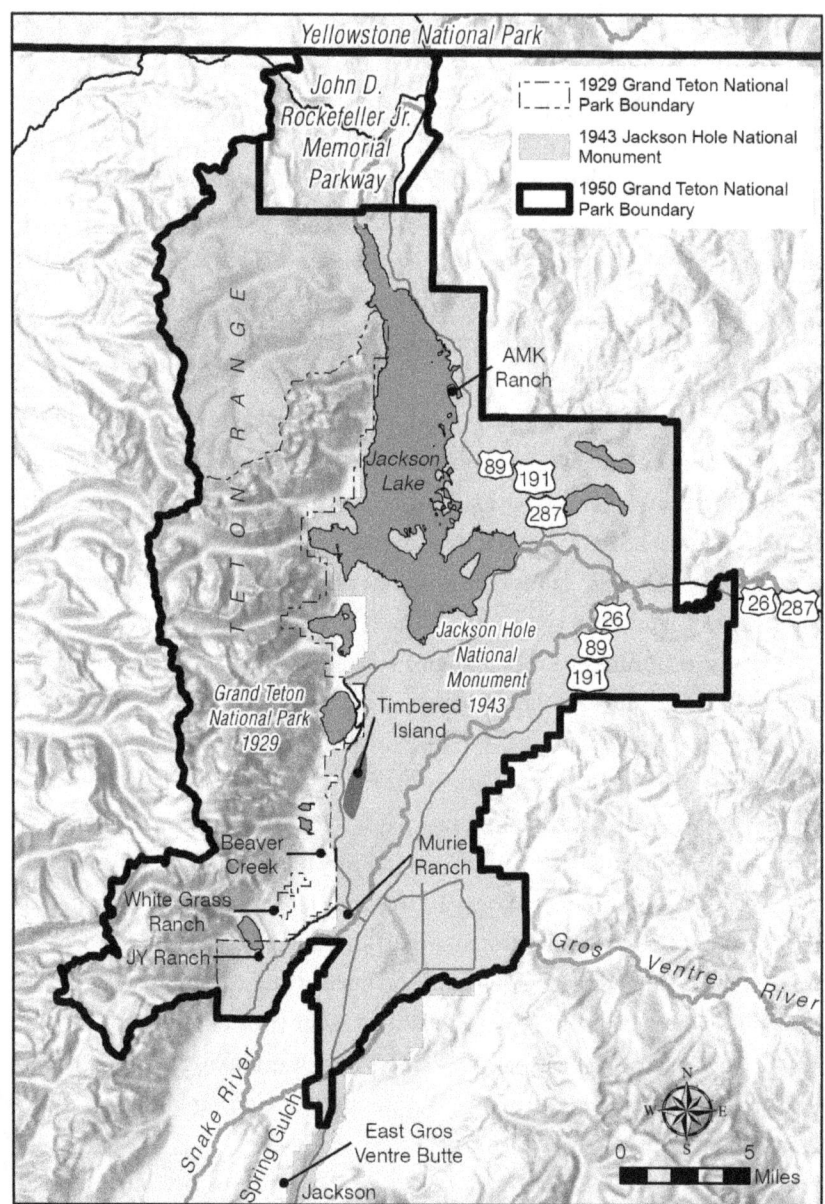

Yellowstone National Park

John D.
Rockefeller Jr.
Memorial
Parkway

1929 Grand Teton National
Park Boundary

1943 Jackson Hole National
Monument

1950 Grand Teton National
Park Boundary

T E T O N   R A N G E

AMK
Ranch

Jackson
Lake

89 191

287

Jackson Hole
National
Monument
Timbered 1943
Island

26

89

191

26 287

Grand Teton
National Park
1929

Beaver
Creek

Murie
Ranch

White Grass
Ranch

JY Ranch

Gros Ventre River

Snake River

Spring Gulch

East Gros
Ventre Butte

N
W   E
S

0   5

Miles

Jackson

MAP 2. Boundaries of original 1929 Grand Teton National Park, 1943 Jackson Hole
National Monument, and enlarged 1950 park. Map by Megan A. Smith, EcoConnect
Consulting, Jackson WY.

by presidential proclamation in 1943, expanding park protection beyond the mountain range, though still missing some land around the valley's most popular lakes.

Further complicating jurisdictional issues in the valley was the Teton State Game Preserve, established by the Wyoming state legislature in 1905; it encompassed land between the Teton Mountains on the west, Yellowstone Park on the north, the Continental Divide on the east, and the Buffalo Fork of the Snake River on the south. A portion of the preserve opened in the early 1920s to permit the taking of elk, and the area's boundaries and hunting seasons evolved later as well.[8] The state of Wyoming has a long and proud history of wildlife management and employed professional game managers before Grand Teton National Park existed in any form. Thus they had a well-established presence everywhere outside Yellowstone National Park, which predated Wyoming's 1890 entry into the Union.

Legally, the new Monument was under the control of the Park Service and, in his 1943 and 1944 reports, Superintendent Paul R. Franke expressed concern without referencing action taken for that area and its resources: "The Jackson Hole National Monument situation is acute from a wildlife stand-point . . . management of that area has a direct bearing on animals within the park. Elk, deer, and to some extent moose, depend on Monument land and the Elk Refuge for winter range . . . the Monument north of the Buffalo River and east of the Moran-Yellowstone highway is open for the hunting of elk, mule deer, moose and bear. This area was previously a game preserve."[9] (The Wyoming legislature did not actually repeal the Teton or other state game preserves until 1947.)[10] But Franke's statement should not be read to suggest that he was confused over jurisdiction of the Monument, which overlapped the preserve. Rather, he had been cautioned by then–NPS director Newton B. Drury to not enforce Monument rules and regulations, as opponents had objected to hunting and fishing rules and, in fact, to the national park rangers' presence.[11] Wyoming's senior senator, J. C. O'Mahoney, had immediately challenged the Monument's creation, ensured that no funds were provided to manage it during its first three years, and "warned the Interior Department against promulgating any regulations," especially those that would affect private property rights. This land included 17,000 acres of private land, thirty-three landowners with 7,300 head of cattle, and forty-two others who crossed Monument land with 9,200 livestock to reach summer range. Governor Lester C. Hunt said, "State government is not surrendering any of its prerogatives in the monument area."[12] In May 1944 the State of Wyoming filed a civil suit against Franke, as Monument administrator, and in December Congress voted to abolish the Monument, but the president refused to sign the bill. A U.S. District Court judge dismissed the state's lawsuit in 1945.[13]

While park expansion dominated managers' attention, black bears were regulars around String and Jenny Lakes—ever-popular visitor areas located in good bear habitat, supporting huckleberries and other fruit-producing shrubs from mid to late summer, then and now. Dude ranches and campgrounds had trouble with bears in the trash. Rangers reported annual censuses of animals, although today's biologists avoid using the term "census" to describe always hard-to-count populations of bears. Unlike herds of elk or mule deer, which group up in winter on open sagebrush grasslands or pastures, making them easier for observers to fly over and count, bears seldom pack into the same areas, except at rare concentrated feeding spots like salmon spawning streams in Alaska or the former garbage dumps in Yellowstone. We have no way of knowing how park rangers came up with estimates of Teton Park bear populations in the 1940s, which leads to skepticism about the value of those numbers. Certainly, increasing the numbers of estimated bears from year to year, based on the visibility of tracks and other sign, as they did, is not in modern times viewed as credible, since there is no indication that trained observers made any systematic effort to search for such sign and no way of knowing whether the amount of searching was similar from year to year.

Wyoming game managers also reported on wildlife populations, suggesting that there were more black bears based on an increase in the number of animals reported killed during the war years. In some areas, reports that U.S. Fish and Wildlife Service predator control agents had taken more livestock-killing bears than in prior years also were used as an indicator of population growth—as many as forty bears were killed in 1946 in the Bridger-Teton National Forest alone. (Biologists admitted that increase could have resulted

FIG. 2. Black bear track at Pacific Creek, Grand Teton National Park, by Victor H. Cahalane, October 19, 1944. National Park Service Harpers Ferry Collection 1607 negative #11181.

from more field coverage by department personnel and from better reporting methods.) Although game and fish departments even today sometimes use hunting or trapping success as a means of estimating predator population trends, it is often criticized as a "trailing indicator"—by the time an agency finds that hunters report low levels of success, the animal population could have taken a dangerous downturn.

Outside of Yellowstone, the Wyoming Game and Fish Department thought that numbers of grizzly bears had remained static but provided no estimate.[14] Between 1940 and 1945 Grand Teton staff consistently estimated that ten grizzly bears inhabited the most remote regions of the park.

The unreliable nature of the historical reports and "censuses" prompts skepticism about early estimates of bear populations, although agency concern and a decline in reports from rangers and hunters signaled a declining trend. In 1945 the superintendent reported that spring and fall hunting seasons adjacent to the park had probably reduced the black bear population and that, although reliable observations of grizzlies were lacking, hunters killed two adjacent to the park and "the grizzly bear situation is considered rather critical."[15] Sadly, this statement is not followed up by any other park records for years, save for one last Naturalist's Report from July 1948, in which "the report of a Grizzly bear seen on the slopes of Mt. Hunt in Granite Canyon was given by E. B. Van Houten (fire guard at the Whitegrass [sic] Patrol Cabin). This is the first report of such an observation being made in this section of the park."[16] In examining the historical records, twenty-first-century researchers concluded that grizzlies were absent from the southern portion of the ecosystem by 1940.[17]

In the late 1940s, the nation focused on celebrating the end of World War II, and as the decade came to a close, local public opinion was turning toward acceptance of an expanded Grand Teton National Park. The government had finally accepted John D. Rockefeller Jr.'s donated lands. Agreements were made to permit existing livestock grazing to remain for a time, and for the secretary of the interior (represented by the NPS) and the Wyoming governor (represented by the Game and Fish Department) to jointly conserve and manage the elk herd by permitting hunters to participate in a reduction program as needed. Other parks have had to address effects of game harvest near their boundaries, and although this activity *within* parks had been against Park Service policy since 1918, hunting did and will continue to affect the Jackson Hole parks.[18] With these compromises, on September 14, 1950, the U.S. Congress incorporated additional national forest and former state game preserve lands into the 310,000-acre Grand Teton National Park that exists today.

Controversy over the new park would take years to wane, but meanwhile managers began clarifying the future of a much-expanded area with an interesting mix of jurisdictions and uses not commonly found in the other national park units of the day. Those conditions, as well as the postwar visitor boom that began in the 1950s and grew afterward, would challenge park managers and the native wildlife, including bears, in ways that continue to play out decades later. And it is doubtful whether park staff or others at the time recognized that one of those native species—the grizzly bear—had functionally disappeared from Grand Teton National Park.

4

Boom Times and Missing Grizzlies

THE 1950S–1970S

I'm one of the declining generation of "baby boomers," born in the post–World War II era. When my two siblings and I were quite young, my parents bought a big wall tent, common in those days for families. It was made of thick off-white canvas and had a floor, and it was tall enough to stand in with ample room for four or five people (if some of them were little) to sleep. With that, we made our way from home in Ohio, in a station wagon sans seatbelts, to park or forest campgrounds, first in Michigan or Kentucky, and thankfully we encountered no bears while tenting. By the early 1960s we graduated to a small trailer pulled behind the car, prompted, I am sure, by Mom's insistence that sleeping on the ground and cooking out of a tent, regardless of weather, had worn out its attraction. I was about nine years old when I first saw black bears begging along the roadsides of Great Smoky Mountains National Park in Tennessee and North Carolina. Whether park rangers and signs discouraged it or not, our parents wouldn't let us out of the car to get close to bears, nor would they let us toss food, though we saw plenty of that behavior accompanied by a not-uncommon reaction when food was withdrawn, as a bear swatted at hapless park visitors. In 1966 we increased our geographic range and headed west with growing numbers of other tourists, adding Yellowstone, Glacier, and Grand Teton National Parks to our summer camping destinations. In

Yellowstone, we saw then-ubiquitous black bears begging for food along park roads, and we often heard the clang of trash can lids opened at night by the ever-clever ursids. Though I can clearly recall a challenge to swim in both Yellowstone and Jackson Lakes with my brother (who "won" the contest to see who could last the longest in the frigid waters of both), I remember seeing no grizzlies, only black bears, and none in the Tetons, though the grand scenery made a lifelong impression.

In 1929, 51,500 recreational visitors came to the newly established Grand Teton National Park.[1] Though the number of park guests grew each year in its first decade, it was a modest trend and, as in most parks, it declined during World War II—to a low of just 8,203 in 1943. In the first three years after the war ended, however, visitation grew to match a prewar peak of more than 150,000 people per year and then boomed, topping one million for the first time in 1954. Regrettably, records of bears or bear management from the newly enlarged park's first decade are scant—they may have been lost to the dumpsters of the day (which would not have been "bear-proof"!), so readers can speculate as to what was happening with wildlife and its management.

National Park Service (NPS) rangers traditionally protected people and resources, educated visitors, and provided public information and services. Over time "protection" rangers were distinguished from those that focused on visitor education, who were first called "ranger-naturalists" and are today labeled as "interpreters" of natural and cultural history. Either type of ranger could be involved in surveying or managing wildlife. Most national park units operate under what is termed concurrent federal and state jurisdiction, and cooperation between agencies is prudent, even if sometimes contentious. The state Game and Fish agency was and remains in charge of fish and wildlife on national forests and private lands in Jackson Hole and elsewhere. But due to their previous responsibilities on formerly private or game preserve lands that had been encompassed by the park, Wyoming Game and

Fish Department biologists continued to play a fairly active role in monitoring and protecting resources at Grand Teton National Park. Citizens and students of wildlife management are often informed that each of our united states owns and has the power and responsibility to manage wildlife within its boundaries. Such assertions may have been especially strong in Jackson Hole due to the long controversy over park expansion, and since the park had limited staff, superintendents may have appreciated the assistance of professional biologists and wardens. Through the 1960s and even afterward, while rangers took part in wildlife management, records suggest that state agents were often involved in monitoring, trapping, and moving bears when such actions were needed.

By 1955 Wyoming Game and Fish Department biologists were concerned that both black and grizzly bear populations in the state had been cut in half. They estimated that only fifty grizzlies remained in Wyoming outside of Yellowstone National Park, and under existing laws and regulations, approximately 85,000 people a year had licenses under which they could conceivably hunt those bears—1,700 hunters for each animal. In addition, seventy-four hunters were permitted for each of the estimated 1,120 black bears in the state, and both species were subject to professional predator control on livestock range. According to the state, "Even in the national parks, bears did not enjoy complete protection. In these areas, they had to be docile enough to get along with thousands of tourists, most of them from big cities, or else face deportation or death before a Park Service firing squad." Wyoming Game and Fish Commissioner Lester Bagley cautioned, "We must soon decide whether we wish to exterminate bear because of their role as possible and actual livestock predators or whether we wish to manage them . . . as native game animals worth preserving. Another year or two of indecision and there will be no decision left to make."[2]

Reports of bear incidents, property damage, and associated management actions began to appear, rather consistently in substance if not in form, beginning in 1958.

TABLE I. Property damages by and related management relocations (REL) and removals (REM) of black bears, grizzly bears (in parentheses), and unknown bears (unk), 1929–2023.*

Year	Incidents	Bear REL/REM	Year	Incidents	Bear REL/REM
1929	n.d.**	n.d.	1989	4	1 / 0
1930–39	n.d.	n.d.	1990	4	3 / 1
1940	1+	0 / 1	1991	3+	1 / 1
1941	0	3 / 0	1992	3	2 / 0
1942	0	1 / 0	1993	5–7	2 / 0
1943	0	1 / 0	1994	4, (5)	6 / 0
1944	0	0	1995	1, (12)	3, (1) / 0
1945	0	2 / 0	1996	1, (14)	2, (2) / 1, (2)
1946–57	n.d.	n.d.	1997	7 (3)	1 / 1
1958	11	12 / 0	1998	7	0 / 0
1959	3	3 / 0	1999	4–7, (2–3)	2 / 0
1960	18	18 / 14	2000***	7–9, (1)	1 / 1
1961	0	0	2001	5	0 / 2
1962	0	0	2002	2, 1 unk	0 / 0
1963	9	16 / 2	2003	6, 1 unk	0 / 0
1964	29	6 / 0	2004	2	0 / 1
1965	42	17 / 0	2005	2	0 / 0
1966	39	19 / 2****	2006	2, 2 unk	0 / 1
1967	19	12 / 1	2007	5–10, (1?)	1 / 4
1968	5	8 / 0	2008	2, (1)	(1) / 1
1969	19	4 / 1	2009	2, 1 unk	0 / 0
1970	8	0	2010	5, (3), 3 unk,	1 / 1
1971	3	1 / 0	2011	3	0 / 0
1972	5	7 / 1	2012	5, 2 unk	1 / 0
1973	0	5 / 0	2013	3, (1), 2 unk	(1) / 0
1974	0	1 / 1	2014	6	1 / 2
1975	4	3 / 0	2015	7	0 / 3

1976	27	9 / 1	2016	0	0 / 0
1977	73	16 / 1	2017	1, 1 unk	0 / 1
1978	18	3 / 0	2018	2, (1), 2 unk	0 / 3
1979	56	4 / 0	2019	2, (1), 1 unk	1 / 1
1980	64	9 / 1	2020	1, (1), 1 unk	3 / 1
1981	25	3 / 1	2021	1, (1)	(1) / (1)
1982	7	3 / 0	2022	2	2 / 0
1983	7	2 / 0	2023	6	1/0
1984	0, (1)	1 / 0			
1985	2, (1)	1 / 0			
1986	6, (2)	4, (1) / 0			
1987	22	7 / 0			
1988	8	6 / 0			

* relocation = bear moved to another part of the park or out of the park; removal = bear taken into captivity or euthanized.
**n.d. indicates no data or record found.
***Data on black bears from 2000–2006 comes only from the GTNP computerized database.
****Bears accidentally killed by drug overdose during capture attempts.

One can wish for more reports that included details, such as "on July 13 [1959], a bear went through a sports car in the Glacier Trail Parking area, both doors on the car were bent and a luggage rack was torn off."[3] For 1958–1969, rangers reported averages of more than ten bears relocated from the park and seventeen bear-caused property damages per year. (For two of those years there is no record on file). These were all black bears, 122 of which were moved with 20 removed from the park, far exceeding the number handled in any later decade. Just in 1960, 14 were destroyed for entering tents, buildings, or vehicles; fortunately, there were no related human injuries.[4] No grizzly bears were reported being seen or handled in the park in those years.

These are not small numbers but, as in previous decades, the reports reflect little concern about the risks to bears or to park visitors or residents. Yet that same year, the Washington office of the NPS had issued a memorandum on its nationwide bear management program and guidelines, calling for "uniform and strict application." The goals were to protect or restore wild animal populations, to offer visitors the experience of viewing a bear in the wild "not as it feeds in garbage cans or at the traffic hazard 'bear jams' along the scenic park roads," and to provide reasonable protection against personal injury and property damage. The plan called for prompt removal of garbage and of potentially hazardous and habitual beggar bears, along with revitalized signs, literature, and other means of visitor information (though not as a "scare program.") Headquarters wanted no tolerance of employees or visitors feeding bears. Parks with bear problems were encouraged to develop their own bear-proof garbage containers.[5] Records do not indicate if or how Grand Teton staff responded to this directive.

In 1965, when visitation exceeded 2.5 million people and there were forty-two cases of property damage and seventeen bears trapped in the park, the superintendent's report admitted: "Eighty-five percent of the incidents would not have occurred had foodstuffs been properly stored. The need for bear-proof food storage facilities is particularly acute at the Mountain Guide's Camp and the Cascade Canyon Ranger Station and Trail Crew Camp . . . The Timbered Island and Colter Bay dumps and several of the private dumps in the park continue to attract bears. Garbage cans in the Beaver Creek, Taggart Creek, and Jenny Lake residential areas were frequently overturned and some roadside garbage can scavengering occurred along the Jenny Lake and Teton Park Roads."[6]

Despite high levels of bear-human conflicts and bears being removed from the park or destroyed, staff apparently made no changes save for considering placement of warning signs at the entrances to park campgrounds. In the following year bears again damaged the Cascade Canyon trail crew camp and were attracted to dumps, garbage cans, and park roadsides. Although this was described as "continued perversion of the food habits of the bear population in the area," there is no explanation for why the park then abandoned plans to install bear-proof garbage cans at trouble spots. A record twenty-one bears were relocated or killed, but no changes were made even in providing bear warnings.[7] In my twenty-first-century tenure, the first instance of park workers leaving unsecured bear attractants would have warranted immediate correction and severe discipline, perhaps even termination of the responsible staff.

This nonchalance, by today's reckoning, about bear-human incidents was not contradicted in local newspapers. They reported on valley residents successfully hunting bears, including the occasional grizzly bear outside Yellowstone, and on hunting season dates, numbers of licenses sold, and success rates. Other stories exhibited continual and casual interest in bears around camps and in town in the early 1960s: Two men from Dubois reported that a bear visited their camp, stole fish they had caught, and dug up fifteen cans of beer stored in a snowdrift to keep cold.[8] Residents on East Kelly Street, in the heart of Jackson, were awakened one night by a black bear rummaging in their garbage can.[9] Students of the Moran School, located inside the park, eagerly awaited "the arrival of sweat shirts which . . . bear the name of the school and their emblem, a sharp grizzly bear."[10] It's unclear whether they knew the perilous population status of their mascot.

Not until 1967 does some change appear in the news, reflecting concern for both bears and public safety not only in Jackson Hole but elsewhere in bear country. In early spring of that year, Dr. Frank Craighead, whose family maintained a home inside Grand Teton National Park and who was, with his brother, John, in the midst of their groundbreaking study in Yellowstone, presented his film *The Grizzly, a Threatened Species* as a benefit for the local school athletic program.[11] In June a son of Jackson residents was seriously mauled by a grizzly bear while working near Ashton, Idaho.[12] Following reports of separate instances where grizzly bears killed two nineteen-year-old women in one night in Glacier National Park, the *Jackson Hole Guide* published an editorial promoting "Safety in the Wilderness."[13] Editors noted that visitors often ignored park rules aimed at protecting wildlife and people. And in response to an incident the following year, in which a man (illegally at that time) carrying a firearm in Yellowstone shot and killed a grizzly, newspaper editors expressed concern for the sparse bear population.[14] Most often, such news and commentary made no mention of Grand Teton National Park, reflecting the widespread understanding by then that grizzly bears were effectively absent from Jackson Hole and the Teton Range.

Black bears were mostly newsworthy as the occasional local hunter's prize. The risks related to bear-human proximity would not be taken more seriously until the eventual human injuries occurred, or until public awareness of and concern for the number of bears being removed from parks grew. Late in the decade, after Yellowstone documented that black bears had caused 422 personal injuries over an eight-year period compared to only 20 caused by grizzlies, the Jackson newspaper opined that not only was a "grizzly zoo hardly an answer" but that even black bears, like other wildlife, deserved to live in their "wild, uncaged and natural state." The editors also took other newspapers to task for comparing the "vicious" grizzly to the common "'little brown' . . . or black bear that harms no one" unless they break park rules against feeding or interfere with cubs.[15]

Teton Park managers had to have been influenced by the bear-caused human fatalities in Montana and attention to Yellowstone's grizzlies, prompted by the Craigheads' research and the then-fledgling controversy over closing the garbage dumps where their studies had been focused. (That fascinating story, which had long-lasting implications for American bears in many ways, is well told in Paul Schullery's *The Bears of Yellowstone* and elsewhere.) Also, likely leaking slowly but inexorably out of Washington DC were implications of the Secretary [of the Interior]'s Advisory Committee on Wildlife Management in the National Parks, more commonly known as the "Leopold Report" after the group's chairman, University of California at Berkeley professor A. Starker Leopold. The report, which had long-lasting effects on NPS policies, effectively reoriented national parks away from human-manipulated or artificial presentations of scenery or wildlife and toward more "naturalness." And it promoted research in and for parks, along with trained biologists, to provide the basis for resource management programs.[16]

In 1970 Grand Teton staff prepared their first official bear management plan, a seven-page document that began by adopting the objectives "suggested by the Natural Sciences Advisory Committee." (This group is not further explained in the plan, but it refers to a special team, also chaired by Starker Leopold, commissioned to review the dump closure controversy in Yellowstone.) The program's objectives, following Yellowstone's lead, were:

To maintain populations of grizzly and black bears at levels that are sustainable under natural conditions as part of the native fauna of the park.

To plan the development and use of the park so as to minimize conflicts and unpleasant or dangerous incidents with bears.

To encourage bears to lead their natural lives with minimal interference by humans.

The plan recognized that visitor safety was a primary consideration and stated that "no differentiation is made in this plan between black or grizzly bears since we have few, if any, year-long resident grizzlies," although it called for increased vigilance to preserve grizzlies' presence in the park. Core bear management actions remain largely the same today: Relocation of a "problem" bear could occur multiple times, and removal of bears was a last resort. Staff were to be properly trained and activities should ideally be conducted away from areas of concentrated human use. Bears were to be discouraged from taking up residence in backcountry campsites or in developed areas used by park visitors and/or residents. Staff were to maintain public education and frequent garbage removal, with individual residents' cans kept inside garages or not put outside until pickup day. (This was prior to the development and installation of "bear-resistant" trash dumpsters and cans.) The plan spelled out requirements for reporting bears and bear incidents and for trapping and handling bears.[17] It is not readily apparent that the new bear management plan resulted in fewer instances of bears causing property damage or being relocated, even removed permanently from the park.

The park's annual bear incident and management reports into the 1970s were submitted by district or chief rangers, which is not meant to imply any criticism of the reports' quality. Instead, it reflects a prior lack of biologists focused on bear monitoring and management. Ironically, Grand Teton National Park had more history than some parks of professional scientists on staff, including Adolph Murie, who worked in the park in the 1940s and

1950s, followed by Glen Cole and Douglas Houston in the 1960s, but neither of the latter did significant work on bears. Houston worked on Jackson
elk and moose. Cole was hired by Grand Teton in 1962 but transferred to
Yellowstone in 1967, ostensibly to oversee biological work in both parks.[18]
He eventually moved Houston to study elk there as well, and both names
are absent from bear management records concerning the smaller park. This
reflects the continued focus on ungulates—hooved mammals—in Jackson
Hole, while both the Park Service and the public were attuned to bear management in America's first park and, to some degree, the other more well-
known and frequently visited "bear" national parks—Yosemite, Glacier, and
Great Smoky Mountains.

As visitation to parks across the nation boomed, rangers—long used to
multitasking as educators and protectors of resources and people—faced
more urban types of law enforcement challenges. A 1970 riot in Yosemite
National Park's Stoneman Meadows called service-wide attention to the
need for more law enforcement funding and training. Employees were challenged to handle a greater volume and array of duties, and the NPS began
moving away from assigning park rangers too many general responsibilities,
influenced by the trend toward increased professional specialization. Meanwhile, influential voices worried about the "State of the Parks," calling for
more research and science-based management. Combined with changes
underway in response to the Leopold Report, this necessitated a restructur-
ing of "bureaucratic power" that, according to former NPS historian Richard
West Sellars, had never been vested in biologists or other staff in science-
oriented positions.[19] Beginning in 1974, annual bear management reports for
both Grand Teton and the John D. Rockefeller, Jr. Memorial Parkway, established in 1972, were signed and submitted not by rangers but by resource
manager Robert "Bob" Wood, who appears to be the first biologist assigned
to deal regularly with bear management.

Wildlife observation logs from 1970 and 1973 note up to ten reported
grizzly bear sightings each year in or near northern Grand Teton Park by
seasoned valley residents or employees, from Slim Lawrence and Frank
Galey to ranger Tom Milligan.[20] Of some note in 1971 was the first report of

a "fruitless" attempt to trap and relocate a grizzly bear from the Colter Bay development after two reported "personal disturbances" attributed to one bear.[21] Even though grizzly bears were still mostly out of sight and mind for Teton staff and guests, they became a new formal administrative responsibility when, in 1975, the species was listed under the Endangered Species Act.

Grizzlies were classified as "threatened," meaning that populations were in danger of being classified as "endangered" unless further action was taken. Fewer than 1,000 of the animals were thought to be left in the lower forty-eight states, including in greater Yellowstone and five other parts of the Northwest, especially around Glacier National Park in Montana. Yellowstone area grizzlies were thought to number from 229 to 364 just prior to their listing, which was in response to concerns about high numbers of bears being killed in Yellowstone after closure of their dumps and due to continued legal harvest of bears outside the parks.[22] Another major concern was the continuing loss or fragmentation of habitat, which isolates bear populations from one another—particularly animals in the Yellowstone ecosystem, being farthest from bear-occupied areas in Canada. Listing the species meant that "it is now unlawful to kill, capture, harm, harass, import, or export a grizzly bear anywhere in the lower 48 states, or to sell any parts or products of grizzlies in interstate or foreign commerce."[23] (Originally one exception allowed sport hunting to continue in northwestern Montana as long as no more than twenty-five bears were killed annually by hunters and wildlife managers; the exception was no longer permitted as of 1991.) Agencies recognized that conflicts with humans were central to the grizzly's decline and recovery.

Some evidence suggests Grand Teton Park staff were calmly analytical or even skeptical about their obligation to engage more in bear management. In mid-decade, resource manager Wood responded to a service-wide survey of bear problems and actions, noting that in both the park and the parkway both species were historically and then present, but that grizzly bears were only "rare or infrequent transients." He reported that he and park rangers shared program responsibility for actions such as storing and using traps, dart guns, and immobilization drugs to capture, mark, and relocate bears. In answering a question about whether a Yellowstone-type rule for food storage was warranted nationwide, Wood replied that "if you have a Yellowstone-type problem it could help—we do not have a problem of that magnitude or regularity" in Grand Teton National Park. He believed bear-proof facilities for backcountry campers were unsightly and did not work. He did, however, write that such a regulation would benefit the parkway, where grizzlies were expected to expand their range from Yellowstone.[24]

In a June 1975 memo to the superintendent, the chief of maintenance wrote that he was "totally unaware that a bear problem exists" and questioned spending an additional $25–35,000 to bear-proof large trash containers. He mentioned a 1972 study that stated bear-proofing was unnecessary and asked for information on the history of visitor injuries and bears causing problems in campgrounds or being removed. But he also expressed legitimate concern for garbage collection in concessioner-run sites, at private ranches and inholdings and at residential areas all through the park, concluding that "if the bear problem is that serious, we should . . . provide for bear-proofing in all collection containers" and look at the entire bear management program.[25] It seems that several years passed before there was serious consideration of these suggestions, whether due to bureaucratic inertia or other factors.

There was a relative lull in black bear–caused property damages and associated bear relocations and removals from 1970 to 1975, but the last half of the decade saw a tremendous increase in incidents. Rangers continued to actively participate in bear management by closing trails and backcountry campsites when necessary in the busy summer of 1976, which, as far as I could find, saw the park's first recorded bear-caused human injuries. Also, managers had to report their first dead grizzly as the result of an incident in the parkway. The law establishing the parkway permitted the continuation of hunting (not of then-threatened grizzlies), which had been legal when the land was managed by the U.S. Forest Service. In September both a bear hunter, who happened to be co-owner of Huckleberry Hot Springs Resort in the parkway, and the employee to whom he reported his kill mistakenly identified a grizzly taken near the springs as a black bear, prompting the park to provide a mandatory animal identification course for rangers.[26]

After rather a disastrous year for bear management, then–park super-intendent Bob Kerr directed a team of five staff, including Bob Wood, to update the bear management program because of numerous challenges, which he did not expect to decline. Kerr did not specifically mention the threatened species listing. But to this former bureaucrat accustomed to dry official internal and external correspondence, Kerr's memo of late October of 1976 was refreshingly thoughtful and candid with clarity of direction that, on the few occasions I experienced something similar in my forty-year career, was seldom put in writing. He directed his committee to consider whether garbage removal was adequate or whether more pickups and bear-proof cans were needed; whether staff had enough bear traps, drugs, and trained personnel; and whether they were ready to work with grizzlies. He spoke to the "emotional part of a bear management program"—destruction of a bear by whom, how, and for what infractions: "Do we destroy an animal after so much property damage; after one bite or scratch; after it has returned so many times after being trapped?" And under the header of Public Relations he wrote, "we need to get our act together and be perfectly open with the community (public) as to what our program is and why."[27]

The committee completed a revised bear management plan by April 1977, ahead of the visitor season, tweaking policies and procedures to clarify that they covered both the park and the parkway. Bear trapping was to occur only if "the offending bear" was known to be present. Bears trapped for prop-erty damage were to be marked and released at least as a first effort, whereas bears trapped for causing personal injury were to be destroyed. Grizzly bear captures and transplants would be coordinated with the Interagency Griz-zly Bear Study Team and Yellowstone. The plan, which was still not lengthy, listed twelve areas in need of attention, aiming to have bear-proof garbage cans or dumpsters at developed areas, at trailheads and roadside picnic areas, along Grassy Lake Road in the parkway, and at the Beaver Creek and Moose administrative and housing areas. (This work was years in the making. While visiting a biologist friend who lived in Moose in 1986, we detoured from our Saturday evening fun while he checked the area and, with chagrin, removed a standard, distinctly *not* bear-resistant trash can from the visitor center porch and locked it inside.) Concessioners were to maintain bear-proof enclosures where necessary—Flagg Ranch and the Huckleberry Hot Springs concession were highlighted. Any remaining dumps in the park were to be closed. Various responsibilities were assigned to maintenance supervisors and district rangers; resource manager Wood was to train personnel in bear trapping and immobilization procedures. The updated plan confidently predicted that "a continuing, aggressive approach to removal of unnatural attractants and prompt bear management should solve our problems."[28]

Would that it were so, then or now. In 1977 the park experienced an uptick in bear activity, with three human injuries, fortunately all "of a minor nature." Someone sighted a black bear at the Lower Saddle, the primary approach to climbing the Grand Teton, at an elevation of 11,600 feet above sea level. With a record seventy-three instances of bear-related property damage totaling more than $5000, more than $2000 of which was done to vehicles, the park relocated sixteen bears, and one was intentionally killed by park staff.[29] While the number of bear-caused human injuries and property damages and bears relocated or removed did decline in the next few years, levels would not have been termed insignificant in effect on either people or bears. In the 1970s, staff recorded 203 instances of property damage, 45 black bears being relocated, and 0–2 verified grizzly sightings each year, although grizzly bears were still declining in Yellowstone and elsewhere in Wyoming. Changing business as usual with regard to bear management in Grand Teton took quite some time, new plan or not.

5

Making the List

THE 1980S–1990S

I was a college undergraduate in 1975, when grizzly bears were protected by the Endangered Species Act. As a child, I had grown up with family Sunday evenings around our freestanding color television console, watching Mutual of Omaha's Wild Kingdom and National Geographic specials. Memorable among the latter was the 1967 Grizzly! program, featuring the Craigheads studying the bears in Yellowstone National Park. The pioneering biologists were the first to use radiotelemetry to track wildlife and designed the first radio collars to fit large carnivores. (I recall footage of them working off-road in the Hayden Valley out of their station wagon, a model not unlike the one our family used to haul our camper trailer for summer vacations, and thinking, could they not even get a pickup truck for their field work?) Their research was instrumental to the U.S. Fish and Wildlife Service's determination of the bears' insecure status.

At the University of Wyoming in Laramie, I was thrilled to hear Frank Craighead talk about their study, and though it would be a half dozen or more years before I first saw a grizzly bear in the wild, I know I felt good knowing that Americans recognized the imperative to save them. Two decades later, I was working in Yellowstone on grizzly bear conservation and public education among other things. One day a colleague came from the backcountry into park headquarters carrying a piece of thick, twisted rope, 10–12 inches long, though no longer a complete circle. It still held a small metal box and remnants of strong tape, similar to black electrical or duct tape. We speculated that the rope was an old Craighead-era grizzly bear radio collar, and it prompted discussion about the evolution of research and of efforts made toward grizzly bear recovery in our lifetimes.

Whether managers and park users promptly recognized grizzly bears as an existing and future challenge or not, a new day was on the horizon. In a 1980 newspaper interview, incoming superintendent Jack Stark said, "Will there be grizzly bears in Grand Teton National Park in 1990? I certainly hope so."[1] A 1982 report on grizzly bear distribution in Jackson Hole between 1936 and 1979 found thirty-two credible reports of bears concentrated along the northern shores of Jackson Lake and in the Rockefeller Parkway, which may have been associated with the then-closed Flagg Ranch dump and the nearby Snake River Campground.[2] Only two of the reports were prior to 1960; the other reports occurred in almost every year from 1968 on. The authors also noted three dens, one of which was confirmed by an Interagency Grizzly Bear Study Team telemetry flight in 1979. Although most of the reported sightings were of single bears, observers had reported one sighting of two subadult bears and another of a sow (female) with two yearlings—an indication of resident bears, even if few people knew of their presence.

Designating the grizzly bear as a threatened species prompted a flurry of federal government action. One of the first responsibilities of the U.S. Fish and Wildlife Service, as the lead agency overseeing listed terrestrial species, was to produce a recovery plan. Agency managers and the public engaged in sometimes fraught debates over designation of critical grizzly bear habitat, though it was not then required. In 1976 the Fish and Wildlife Service had begun to outline what and where that should be—areas that meet the animals' needs for nutrition, movements, reproduction, cub-rearing, growth, and natural behaviors. But early draft maps prompted an outpouring of opposition from hunters, ranchers, outfitters, snowmobilers, and the timber industry even though such designation would not prohibit multiple

FIG. 3. Large adult grizzly bear. Photo by Gary M. Pollock.

forest uses.[3] U.S. senators Cliff Hansen and Gale McGee appeared at public meetings in Cody and Jackson to hear citizens' concerns. The Wyoming Game and Fish Department asserted that only Yellowstone National Park should be termed critical for the ecosystem's grizzlies, as the state had the jurisdiction and ability to manage the bears effectively in other areas. Grand Teton National Park endorsed the concept of critical habitat, although then-superintendent Kerr did not expect much resulting change in park management.[4] John Craighead, in an effort financially supported by the Bureau of Land Management, the National Park Service, the Forest Service, and the Fish and Wildlife Service and based on data from observations, movements of marked grizzlies, and an analysis of their spatial and food needs, outlined an area including and surrounding Yellowstone, including the northern half of Grand Teton Park (and the Rockefeller Parkway, although it was not labeled on the map).[5] The first Grizzly Bear Recovery Plan, in 1982, avoided the contentious "critical" terminology but showed the parkway and northwest corner of Grand Teton National Park as part of then-occupied habitat, mirroring the reported bear sightings. Although Congress soon thereafter amended the Endangered Species Act to require designation of critical habitat, it was not retroactive and, in the end, public controversy prevented it from ever being declared for grizzly bears in the lower forty-eight states.[6]

When the government prepared an updated recovery plan in 1993, grizzlies occupied habitat throughout the parkway and about one-third of northern Grand Teton National Park. Thus, the grizzly bear recovery zone boundary, then and now, follows Grand Teton's North Park Road (shown on some maps as U.S. Highway 287/191/89) from the eastern park border through Moran Junction and Entrance Station toward Yellowstone and along the shoreline of Jackson Lake from about Lizard Creek Campground— where the road moves closest to, then eastward away from, the water—up to the Rockefeller Parkway's southern border. Then the boundary turns back south almost to Mount Moran, which is outside the recovery zone, and west to the Teton crest. (See maps 3 and 4, chapters 9 and 10.) South of Jackson Lake, no portion of Grand Teton National Park is included in the recovery zone, although that does not preclude the park from taking actions on behalf of grizzly bears.

The original recovery plan required all agencies to address illegal and accidental deaths of bears, such as those caused by black bears being mistakenly identified as grizzlies and hunted or trapped. This was also intended to strictly govern agency grizzly control actions. And the plan aimed to reduce or eliminate recreational activities, mining, timber operations, and livestock grazing that limited the bear population.[7] The U.S. Forest Service led the effort to produce *Interagency Grizzly Bear Guidelines,* which outlined population and habitat conditions as well as direction for five different "management situations."[8] In Management Situation 1—areas key to grizzly bears' survival—management decisions should favor the bears when habitat and other land uses compete. Each federal land management unit in the recovery zone mapped its "MS" lines, and 95,373 acres of Grand Teton National Park and the John D. Rockefeller, Jr. Memorial Parkway are in MS 1. The park and parkway assigned 2,355 acres of facilities—including Flagg Ranch

Village, the Snake River Campground, and the Snake River gravel pit, all just off the road connecting Grand Teton and Yellowstone Parks—to Management Situation 3, where the animals' presence is possible but infrequent or not desirable; bear-human conflict minimization is a high priority, and bears may be actively discouraged from hanging around. Ironically, all of Yellowstone National Park was classified as MS 1, despite having developed campgrounds and lodging areas inside its borders. No portions of either park or the parkway—even areas still grazed by livestock—were defined as MS 2, where the grizzly bear is an important, but not primary, use of the area and in which managers would control nuisance grizzlies. Since only Grand Teton Park's small Lizard Creek Campground, and none of the high-use areas, was included in the grizzly bear recovery zone, management situations have seldom seemed pertinent and a map depicting them is hard to find.

A second topic addressed by the guidelines was control of those "nuisances," defined as bears that: a) caused significant depredation to lawfully present livestock or became food conditioned; b) displayed aggressive as opposed to defensive behavior to humans, posing a demonstrable threat to human safety or causing a minor human injury; or c) caused substantial human injury or death. For each case, depending on the sex and age class of the bear(s), whether it was a first or repeat offense by the same animal, and other circumstances (such as, "attractants were removed or stored so as not to attract grizzlies"), the guidelines suggest relocation or removal of the involved grizzly. The guidelines have been used in addressing livestock conflicts in the national forests and in Grand Teton National Park.

Beginning in the 1980s, Teton Park's superintendent joined other state and federal managers at semiannual meetings of the (*take a deep breath— even the participants abbreviate it as "YES"*) Yellowstone Ecosystem Subcommittee of the Interagency Grizzly Bear Committee—the umbrella group of federal, regional, and state directors of agencies responsible for grizzly bear recovery in the United States—to discuss collaborative efforts to protect bears and habitat. Early on, major discussions addressed bear attacks on people, including two fatal incidents in Yellowstone National Park; proposals to hunt bears outside the parks and, conversely, to increase their protected status to "endangered"; and whether to provide supplemental food to grizzlies. Such issues may have seemed distant to the Grand Teton National Park superintendent who first reported property damage in 1984, when a yearling female (#121) that had been relocated from Yellowstone obtained food from a cooler the Flagg Ranch manager had left outside overnight.[9] But regardless of Superintendent Kerr's clear direction in the mid-1970s and the more recent incident, the park as a whole appeared slow to acknowledge that bear management could, even should, be improved. That summer, visitors complained about non-bear-proof trash containers at Colter Bay and north to Grassy Lake Road. Some of them had attracted at least one black bear, which biologists removed. A park interpreter called the park's failure to follow its own seven-year-old bear management plan "hypocritical" and "reprehensible."[10]

Staff had to trap their first nuisance grizzly, a yearling male that got into a bird feeder and garbage near Signal Mountain Lodge in October of 1986.[11] In 1988 alone the park received ten reports of grizzly bears, fortunately without incident.[12] In 1989, sightings of grizzlies increased and included a female with at least one cub.[13] A news feature on grizzly bear #139, captured in Island Park, Idaho, and released well east of the parkway, displayed that big male bear's movements in late 1987 and the summer of 1988. The bear roamed, in typically nonlinear fashion, the east side of the valley until he returned north through Grand Teton National Park to winter in southeastern Yellowstone. Out of his den the following spring, he wandered past Flagg Ranch and Old Faithful then found his way to the northwest corner of Yellowstone by June.[14] Such long-distance movements are not uncommon for adult male, sometimes called boar, grizzlies.

These observations and conflict reports of grizzlies getting into bird feeders and unsecured food, at park-managed facilities such as the Brinkerhoff cabin on Jackson Lake, signaled that the long-absent bears, of all sexes and ages, were expanding into new or recently unoccupied habitat—and that there remained work to do in reducing bear attractants around park facilities. Workers across the park notably stepped up their responses. Echoing concerns from a dozen years before, in a 1987 memo to the chief of maintenance, a biologist estimated a shortage of two hundred bear-proof trash cans as well as inoperable and insufficient numbers of dumpsters.[15] Two years later, the chief ranger directed his staff to inventory sites that needed bear poles and food boxes in the front- and backcountry. They found that only 28–47 percent met campers' needs and itemized $323,000 needed over the next five years to more effectively secure human food from bears.[16] In 1990, rangers listed fifteen more areas with a $68,840 price tag to fix trash cans without bear-resistant tops at the Oxbow picnic area and elsewhere, prioritizing sites in Situation 1 habitat.[17] Resource manager Bob Wood had retired after helping initiate significant progress, although it's unclear if the work began that year.

It may have been persistence by Wood's replacement, new wildlife biologist Steve Cain, that prompted meetings of senior managers in December 1990 and 1991 and resulted in more progress toward "bear-proofing." The park sign shop was directed to produce bear-education signs for Jenny Lake and the Death Canyon backcountry camping zone. Maintenance workers aimed to buy and install seventy-five new bear boxes at Jackson and Leigh Lake shoreline sites and at Lizard Creek and Colter Bay campgrounds. At the time, a four-yard dumpster cost $1,000 plus shipping, a bear box $85, and a bear-proof trash can $55. (For comparison in 2023, a dumpster cost $3,800–$5,700, a bear box $1,400 and up, and a bear-resistant trash can upward of $1,500.) Even then, this effort doubtless required cost-sharing and use of various accounts—what clever managers call creative financing. On a copy of meeting notes announcing a 1991 special initiative to provide $16,582 for bear-"proof" facilities, the superintendent complimented Cain's good work, but he cautioned that "$7,500 is all that's in the Bank—there's no gold card for overruns."[18]

By then, Grand Teton had replaced the standard annual bear incident and management reports, filed through the regional office in various forms for the previous three decades, with detailed narratives of the increased attention paid to bear country education as well as securing bear attractants. The park posted an attention-getting "Who Killed the Bear?" flyer after having to move a black bear from the popular Hidden Falls area. Rangers strived to patrol campgrounds nightly to talk to campers, to enforce food storage regulations, and to file more reports of bear sightings and "case incidents," though, if completed, most of the latter have not made it into the park archives.

Employees were especially alert for grizzly bear attractants each autumn, associated with hunters in and adjacent to the park. Participants in the elk reduction program could, at that time, camp overnight in the park. They set up at the end of Pilgrim Creek Road and near the start of the Pacific Creek Road, both side routes off the main highway between Moran and Colter Bay. The park installed tall carcass-storage racks and hanging poles at the hunt camps. Nevertheless, a female grizzly and her cub-of-the-year got into a sliding-door dumpster near the new meat racks at Pilgrim Creek camp in September 1992, reminding staff that continued vigilance and bear-proofing *all* facilities was critical; the dumpsters were replaced with receptacles with more secure latches.[19] (After the early 2000s, use of the camps declined due to shifts in elk and hunter distribution as well as users' tendency to overnight closer to amenities found in Jackson.)

The park also made progress toward reducing other sources of bear conflict as funding became available. In Grand Teton's annual update to the managers' subcommittee, rangers reported that while patrolling to look for unsecured bear attractants, they would ensure that gates to the fenced sewage lagoons near Colter Bay were closed.[20] In the same report, neighboring Yellowstone listed their need to build bear-resistant fencing around several of its sewage treatment plants, reinforcing the oft-forgotten attractant presented by sludge, which grizzly bears occasionally investigated and rolled around in. (*One is so tempted to say, you can't make this sh— up!*) It's funny how these things can circle back around years later.

The last decade of the twentieth century passed with continual progress in making Grand Teton less conflict prone for black bears and, none too soon, for the grizzlies that had by then begun their own recovery in Jackson Hole. Sightings of grizzly bears or their tracks, by at least a few observers, had become an annual occurrence in the park and parkway. The national forests, the parks, and the U.S. Fish and Wildlife Service had plans and guidelines for grizzly bear management or were putting them in place. Agency personnel were expanding their human networks. I was fortunate to have attended the grizzly bear managers' meetings (among other support staff, sitting in what we called the cheap seats, behind both our bosses and interested citizens) for more than thirty years beginning in 1985 and experienced firsthand the benefits of cocktail hour conversations that enabled increased interagency trust and cooperation, leading to bears being relocated across state and jurisdictional borders. Research was providing more and better information on how grizzlies lived in greater Yellowstone, offering insight on how humans could better live with them. Park managers were awakening to the presence of grizzly bears in the Tetons and to the efforts needed to manage them. Across the ecosystem there was increased hope for saving the threatened species—the USFWS recovery coordinator predicted that grizzlies would be "delisted" from Endangered Species Act protection within the following decade.[21] You've got to love an optimist.

6

The Life of a Bear

One May evening, my husband and I were enjoying a drive in the park and, just before dusk, came upon a small group of cars—a bear jam—at Jackson Lake Junction. We saw one, then two grizzlies grazing quite close to the road between the pavement and the tall willows lining the edge of the Willow Flats. The bears were both good-sized adults, though one was larger than the other, and it soon became clear that the animals were a courting pair. The larger male bear occasionally tried to sidle up to the smaller female and sniff at her rump, though she appeared disinterested. Darkness fell before we saw any more obvious mating behavior. It reminded me of earlier instances when I'd chanced to see bears mating, and it was not as exciting as it may sound, especially at a distance, since it looked like the bears just sat down next to each other for twenty minutes or so. Once, watching in my ranger regalia as two black bears mated a short distance off a park road, a visitor asked, "How old are they?" and I admit I quipped, "Old enough!" (I went on to explain at what age that would be for bears.)

Bears are not what biologists call a highly social species. Multiple bears can be seen at one time when food is not limited—picture images you may have seen of dozens of bears in proximity to each other, feeding on spawning salmon in Alaska's Brooks and McNeil Rivers, or at Yellowstone's garbage dumps in their heyday—but this is not typical in most habitats. To the contrary, bears can be quite dangerous to one another, with males fighting for food or mates. While not often observed, these battles are evidenced by the large, gaping bite wounds boars can leave on each other, seen by few but those who trap them. Like other predators, adult male grizzly and black bears can be especially dangerous to cubs, and occasionally other-age bears of either sex kill their own kind or a different species of bear.[1] If you see bears in proximity to each other, they are likely related, or it is mating season.

Bears in the ecosystem pair off between May and early to mid-July. Male grizzlies can be attracted from miles away to the scent of a female bear in estrus, which can last for several weeks. Black bears are on average slightly younger than grizzlies at first motherhood, and females most often keep their cubs with them only one winter after their birth, whereas grizzlies commonly keep their cubs through their second full winter after emerging from hibernation. A female that still has cubs in tow, a potential mate, or even both may chase off her youngsters to focus on breeding. Both sexes of bears can be promiscuous, mating with more than one member of the opposite sex during the same season. With the increased use of DNA analysis from samples of bear hair, blood, or even scat, biologists know that cubs in the same litter can be sired by different fathers. On average, female grizzly bears in greater Yellowstone produce their first cubs near the age of six, although it can happen between the ages of four and seven, occasionally even later. Litters can be one to five cubs—two is most common, but biologists caution that they seldom know how many cubs are actually born, deep inside the animals' dens, in January or February, compared to how many emerge and survive until watchers may see them in late spring.[2] Observations suggest that black bears in the Tetons most often have two cubs, although in late 2023 a local photographer driving over Teton Pass snapped a shot of a black bear mother with five young cubs, which may have been a record for Wyoming.[3]

Despite the time lapse between mating season and the birth of cubs, mothers are not pregnant for eight to nine months; they undergo delayed implantation of fertilized embryos until about November. Research on captive brown bears found that if a female is in sufficiently good condition, meaning she has accumulated at least 20 percent body fat by the time she dens, fetal development occurs in six to eight weeks.[4] If not, the animal fails to bear young. Hibernating bears forgo eating, drinking, urinating, and defecating in their denning months, though females do produce milk and nurse their cubs. Entry into winter dens, often small spaces excavated under tree roots, appears to be prompted by the lack of food availability and weather conditions. Male bears enter dens later and emerge beginning in March, earlier than females without cubs, which move to lower, snow-free elevations soon after leaving their winter homes. Mothers and cubs typically restrict their movements for several more weeks and often are not seen until late April or May.[5]

Bear cubs weigh less than a pound at birth, and ten to fifteen pounds by the time they begin to move about with their mother as the snow melts. Fully grown male grizzlies in the ecosystem average 457 pounds by late autumn, and females just under 300 pounds as they prepare to enter their dens. Black bear females average half the weight of grizzlies, and males about 260 pounds, though published data on the size of black bears in the Tetons is limited. Bears do not mark and actively defend a territory, as does a wolf pack or a red squirrel, but they do show affinity for the same home range— the area in which an individual bear spends its time over much of its life. In livestock grazing areas of northern Jackson Hole, black bear home ranges documented in the 1990s averaged ninety square miles.[6] Their home ranges were considerably smaller in the southern Tetons, averaging twenty-five square miles for males and only eight square miles for females.[7] By comparison, a large grizzly male's home range averages 154 square miles and can span well beyond the length of Grand Teton National Park.[8] A female grizzly bear's range averages more than sixty square miles, but both species expand and contract their home range from year to year based on whether cubs accompany a mother bear. In their first year of life, the cubs' small size limits their ability to keep up with an adult female moving long distances. As cubs grow into year two or three, they can and do move farther as their parent seeks to teach them all that she knows about the area. Logically, large litters of bigger cubs need the largest area full of high-energy food to satisfy mother and growing brood.

FIG. 4. Two young grizzly cubs playing. Photo by Gary M. Pollock.

The impetus to find food and put on weight only heightens as summer fades to fall, plant foods dry up or become covered with snow, and some mammals or insects that are part of their diet are less available—or come at more risk to bears; think hunter-killed carcasses. For these reasons, bear managers typically steel themselves for an increase in bear-human conflicts and associated management actions such as trapping, translocation, or even removal of bears in the hyperphagic autumn, when bears are driven to consume the maximum possible calories in preparation for winter.

Once young bears leave their mothers, male offspring are more likely than their sisters to wander a considerable distance seeking their own homes, which can take a while to establish. During that time, a subadult bear (akin to a teenage human) may encounter considerable competition and be challenged to succeed in areas where larger bears already live. Thus, subadult male grizzlies have the largest home ranges in the ecosystem.[9] Female grizzlies are known to tolerate their daughters overlapping their own home range, which has resulted in the visible increase in the numbers of mothers and cubs seen by both biologists and park visitors to the ecosystem in recent decades. This occurs where it's not so easily seen, in forests and backcountry areas, but has played out in full view of an increasingly watchful public along Teton park roadways in the twenty-first century. Although grizzly bear #399 was not the first reproductive mother grizzly in Jackson Hole since the 1940s, she was the first most people chanced to see. As the years passed since she appeared with cubs in 2006, wildlife watchers looked for not just her but one of her oldest daughters, #610, then subsequent offspring from both well-known females, all occupying, at least for a time, the central portion of Grand Teton National Park and adjacent lands.

Since the earliest days of the Interagency Grizzly Bear Study Team, biologists have focused on marking and monitoring females and cubs. Although the ecosystem's grizzlies are 50 percent males and 50 percent females, the latter drive the population trend. After the act of conception, male bears do not contribute to raising cubs that can grow up, live on their own, and eventually reproduce. For this reason, not only monitoring but management efforts have been focused on keeping female grizzlies alive and, as much as possible, out of conflict with humans. The biology holds true for black bears as well, though there is no indication that their population levels have declined or are threatened. During decades-long efforts to understand and recover the ecosystem's grizzly bear population, there has seldom been time, money, or impetus—since the two species are similar in many ways—to study black bears, thus there is much less quantifiable information on the smaller species.

Black and grizzly bears do exhibit differences in when and where they spend their time. The smaller black bear is more of a forest creature, easily able to climb trees when needed, and likely seeks the safety of tree cover to lessen its risk from the larger bear species. Grizzlies once occupied a much larger part of the United States than they do today, having been seen often in the tree-sparse Great Plains as the Lewis and Clark expedition moved up the Missouri River toward the Rocky Mountains. (A fun read of those historical accounts is Paul Schullery's *Lewis and Clark Among the Grizzlies*.)[10] The hump on a grizzly bear, often noted as a key feature distinguishing one from a black bear, is a large shoulder muscle developed from digging varied grassland foods. Grizzlies did and do today make more use of open grassland habitats. Although specific studies show some differences, black bears are generally termed "diurnal"—more active during the daytime hours—and female grizzlies more "crepuscular"—most active around dawn and dusk. Male grizzlies are nocturnal—most active at night—using open habitats while black bears rest under forested cover.[11]

Bears are known to be quite curious, such as when they test campers' tents, barbecue grills, or food storage containers. The occasional observer chances to see them engage in what appears to be playful behavior, whether purposeful or not, investigating things from photographers' tripods to anglers' boats. I once received a call from my longtime friend and colleague Dr. Robert "Bob" Smith, now emeritus professor of geophysics at the University of Utah, who grew up in Jackson Hole and still returns often as a part-time resident of the valley. Bob and his research partners had long maintained a series of seismic and GPS stations across Grand Teton and Yellowstone to record and transmit ground movements in the geologically active region, including along the Teton fault, which runs just above the base of the Teton Range. As chief of park science, I approved proposals for research data collection and monitoring equipment installed temporarily or permanently in the park or parkway. One of the long-term stations, located not far west of Jenny Lake above a small waterhole called Moose Pond, had gone off-line, and Bob wanted to send someone in to repair it. Rather fortuitously, that same week a park visitor taking a break while hiking toward Cascade Canyon had looked across the hillside to see a black bear investigating the GPS station. The bear pushed and prodded the small metal tripod holding a satellite dish that powered the monitor, finally knocking the equipment over; part of the unit fell several hundred feet down the mountain toward the pond. The visitor's pictures of the bear's deeds got back to Bob, who shared them with me. We had a good laugh about the bear that foiled the seismic station, which to our knowledge had not happened before.

And bears display a remarkable ability to learn what human observers might call both good and bad lessons about how to survive in their world. An experienced mother bear will spend the time with her young teaching them what and where to forage—ideally away from human roads, housing units, and other temptations that can ultimately cause trouble for both people and bears. As far as we can discern, bears eat to live and live to eat, being on a nearly constant search for food, which comes in wide variety in the Tetons and elsewhere in the ecosystem. It has been so in nature for a very long time. But for a while, the human presence in Jackson Hole influenced bear diets, distribution, and behavior in a way that has, thankfully, become a thing of the past.

7

Dump Days

Lady Bird Johnson spent several days on the JY Ranch while her husband, Lyndon B. Johnson, was the thirty-sixth U.S. President, and one of her daily diary entries noted: 3:30 p.m. Mr. and Mrs. Rock. + [Liz?] toured Ranch, visited Snake River, fishing lake, garbage dump to see grizzleys [sic].[1]

There is well-documented history of Yellowstone Park dumps and associated bear activity, particularly at Rabbit Creek, north of the Old Faithful development; at Trout Creek in Hayden Valley, where the Craigheads centered their study of grizzly bears from 1959 to 1971; and at Otter Creek just south of Canyon Village, where park visitors could sit on bleachers, in a form of grandstand, overlooking a bear-feeding platform until the 1940s.[2] In contrast, there's a paucity of documentation on how frequently bears found dumps in the shadow of the Tetons. The obituary of an early valley homesteader mentioned that in her honeymoon summer of 1931, a grizzly bear came to the garbage pit daily near her home at what was then the Forest Service's Arizona Ranger Station, on Arizona Creek near Jackson Lake.[3] Government facilities obviously had to dispose of their own trash, but one of the noticeable differences between Yellowstone and Grand Teton National Park is the historical and, to a small degree, remaining presence of dude ranches with associated garbage to handle. In the days before and into the

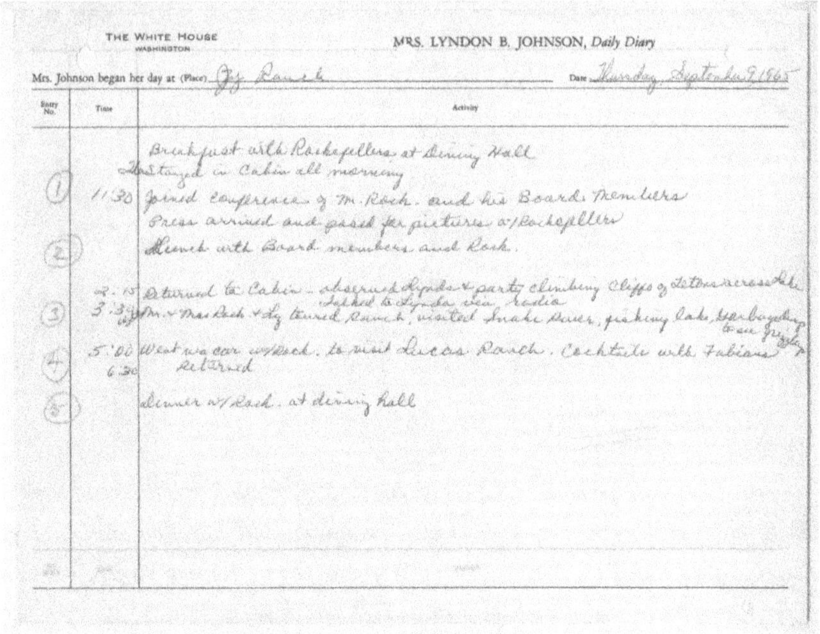
THE WHITE HOUSE
WASHINGTON

MRS. LYNDON B. JOHNSON, *Daily Diary*

Mrs. Johnson began her day at (Place) JY Ranch   Date Thursday, September 9, 1965

Entry No.	Time	Activity
		Breakfast with Rockefellers at Dining Hall
		Stayed in Cabin all morning
(1)	11:30	Joined conference of Mr. Rock. and his Board Members
		Press arrived and posed for pictures w/Rockefellers
(2)		Lunch with Board members and Rock.
	2:15	Returned to Cabin – observed Lynda & party climbing cliffs of Tetons across lake. Talked to Lynda via radio
(3)	3:30	Mr. & Mrs. Rock. & Ly toured Ranch, visited Snake River, fishing lake, garbage dump to see grizzly
(4)	5:00 6:30	Went via car w/Rock. to visit Lucas Ranch. Cocktails with Fabians & Returned
(5)		Dinner w/Rock. at dining hall

FIG. 5. Lady Bird Johnson's daily diary entry for September 9, 1965, National Archives Collection LBJ-PCTJWHD.

1970s, private ranches disposed of refuse wherever they thought convenient and just far enough from where it was generated to be practical. Surely, I'd thought, someone in early Jackson Hole saw bears, even grizzly bears, at a garbage dump—it could not have happened solely within the boundaries of Yellowstone. And here in Lady Bird Johnson's personal diary was an indication that bear watching occurred in Grand Teton National Park, on the valley's oldest dude ranch, and by the then–First Lady of the United States! Somewhat regrettably, this teaser was deflated in a detailed transcript of an oral history interview in which Lady Bird said that day

> was the best of my days in the Tetons . . . A little after 11 I walked over to join the Rockefellers in the main living room at the Lodge, where he is briefing the press over coffee and doughnuts. How well he does

it!... I stayed at the Main Lodge after the press left to have lunch with Laurance and Mary Rockefeller and all of his Board Members, a sort of amalgam of conservationists from many groups and from the various Rockefeller Foundations... I didn't yearn for a nap when Laurance said that he and Mrs. Rockefeller would like to show me the ranch... Laurance driving, we toured the JY, stopped on the banks of the Snake River... Too swift to tame, and very dangerous to fall into... We saw their fishing lake, which they had stocked with trout... we could see the fish perfectly—speckled, about 10 or 12 inches long, their delicious brothers we had eaten that day for lunch. Mary Rockefeller and I had already walked down to the corral that morning, but there was one more place they wanted me to see.

Of all things, the garbage dump! Because it's there where the grizzlies come almost every day to have a fine meal. We sat for about 20 minutes. The quiet forest came alive. We saw the busy little squirrels, birds that sound like bluejays, but a quieter gray color, but NO grizzlies.[4]

The JY Ranch, reached by traveling the Moose-Wilson Road in the southwestern corner of the park, was established in 1906 by Louis Joy who, ironically, staked his claim under the Desert Land Act despite its being nothing like desert, lying mostly along the forested edge of a lake at the base of the Teton Range. Ownership passed in 1920 to Henry S. A. Stewart, under whose leadership it grew to be the largest guest ranch in the valley. It ceased being open to paying dudes after Stewart sold to John D. Rockefeller Jr.'s company in 1932, and the property was a family retreat for more than seven decades. Larry Rockefeller—son of Laurance S. Rockefeller, who hosted Lady Bird Johnson at the ranch and later donated the property to the park—was very gracious in sharing his and his compatriots' memories of the JY when I reached out to him. Larry recalled spending "an annual week or two in August, beginning in 1950," where "bears—black bears, never grizzlies—... would come by the kitchen or cabins on occasion, as well as visit the small dump that was [later] closed." The dump was located across Lake Creek from the horse corral, about a quarter mile from the southeast shore of Phelps Lake. He remembered some stories of bears getting into garbage outside the kitchen, including once "when NPS ... moved a problem bear, via a trailer 'trap' in a big tube with a spring door ... [and] our going down an incline to a cabin by the lake, joking, 'There's a bear; there's a bear,' followed suddenly by, 'There IS a bear!!' and a retreat up the slope."[5]

Larry's cousin Steven Rockefeller, who first visited the JY in 1948 and worked there as a wrangler in 1950 and 1952, spent the summer of 1954 working for the park's trail crew in Cascade Canyon. During those years "there were a few black bears ... around the ranch and they periodically were attracted to the garbage cans behind the kitchen." He recalled that Ken Chorley, president of Colonial Williamsburg and an adviser to John D. Rockefeller Jr.

> is reported to have shot a black bear that had broken into the kitchen at the JY and was on top of the dining room table! If that story is true, it probably occurred in the 1940s ... I never saw a grizzly bear in or around the ranch ... Regarding the story about Lady Bird Johnson ... I was at the ranch when she visited in 1965. There were no grizzlies visiting the dump. They were all black bears, and one in particular was very large.[6]

Larry Rockefeller thought the ranch dump had been closed in the 1970s, although park records noted that a subadult black bear was trapped twice at the ranch in 1986 and relocated after "feeding on readily available dog food and garbage."[7] The JY was not unique in having bears frequent its high-quality forested habitat, or in the way bears took advantage of easy eating. Garbage dumps in the park provided multiple such opportunities for half a century.

Just north of the JY is Jackson Hole's third dude ranch, White Grass Ranch, inside the original Grand Teton Park boundary. Lying below Static Peak at the edge of an expansive meadow once irrigated and cultivated for hay or pasture for horses or a few cattle, the ranch buildings can still be seen by driving up the Death Canyon road, which also provides hikers access to a popular trail leading into the canyon. In the late 1940s, according to Jack Huyler:

Louis Leisinger had a real problem: She was big; she was tough; she was ornery; . . . she was used to having her own way. She hung around the White Grass Dude Ranch. She thought she owned the place. She visited the ranch kitchen every night. She, a big, black, sow bear, wreaked havoc with the garbage can just outside the kitchen . . . Well mannered bears dined down at the ranch's garbage dump . . . Perhaps she preferred her food warm. Exasperated by the cleanup called for each morning before the cook could settle down to preparing breakfast for the crew and then the dudes, Marion Hammond, owner of the White Grass, asked Louie Leisinger to do something about it . . . First, he got a heavy steel oil drum and cut one end out of it with a welding torch. Next, he removed the garbage can. He set his oil drum where the garbage can usually stood. Then into his drum he piled enough rocks that it could not easily be tipped over. A load of tasty garbage was dumped on the rocks. Finally, Louie threw the end of a logging chain over a pine limb directly above the can; and to it, with the help of some bailing wire, he fastened a chunk of granite that weighed 60 to 65 pounds. He heaved on his chain until he had the rock suspended only a couple of inches directly above the can. He tied it off there, and all hands waited for dusk. Sure enough, come dark, they heard the old bear snuffling around outside the kitchen, where Louie and some others were hiding quiet as the proverbial mice. The bear approached the can and shoved the rock aside so she could get her head into the garbage, which was too heavy to tip over. Of course, the rock swung back and tapped her firmly on the head. Annoyed, she gave it a harder shove; but no sooner did she stick her nose into the barrel than here came the rock and gave her a pretty good whack. Infuriated, she shoved that rock aside as hard as she could; and, satisfied, stuck her head back into her intended meal. This time, having been pushed as hard as only a bear can push, the rock had built up considerable momentum by the time it came swinging back and cracked the sow such a blow on the skull that for 50 yards on her way out she staggered like a drunk. Never again did that bear bother the White Grass kitchen garbage can; she joined the other bears down at the dump.[8]

FIG. 6. Black bear outside cook cabin at White Grass Ranch, 1950s. Judith Schmitt Album_347. Jackson Hole Historical Society and Museum and National Park Service White Grass Heritage Project. Photo by Judith Schmitt.

The National Park Service acquired White Grass Ranch as a life estate in 1956, and the property was ultimately turned over to the park in 1985. Although some of the buildings were then sold and moved, remaining ranch structures were restored between 2005 and 2016 and serve today as home for a training facility called the Western Center for Historic Preservation. During that restoration, the park was fortunate to station at White Grass a volunteer caretaker, Roger Butterbaugh, who greeted visitors to the site for eight seasons. Roger also compiled photographs and oral histories from former ranch residents, workers, and guests, some of whom were eager to share their memories of time at what has obviously been a special place to them.

Judith Schmitt went to the ranch as a guest from Philadelphia in 1955, and she subsequently worked there until 1960. In her recall, bears were always around the ranch, and although then-owner Frank Galey told staff not to feed the bears, one cook did. Judith lived in a canvas tent with a wooden base, west of the main lodge buildings, and she literally ran into a brown-colored black bear one night while on her way to the restroom. That might have been the year that Frank imposed a "bear curfew" because there were so many of them between the ranch cabins. She remembered a "great dump to which workers would sometimes ride down after dinner in Frank's truck," and they sometimes saw six or eight bears, always black bears but of assorted colors. At that time, Judith said, people fed bears, and other ranches had dumps that attracted bears as well, but any photos she and her friends took were of people, not wildlife.[9]

Karin Gottlieb, another White Grass employee, who first went to the ranch in 1963, went just once to the dump, and although no bears were there that time, "the chore boys saw bears at the dump often when they brought fresh garbage . . . It was all pretty routine." But when she saw one behind the kitchen upon her return, she did "chase the bear until he/she changed course because I was concerned she'd get into the girls' cabin."[10] Fred Herbel, another former ranch wrangler, who went on to become a Wyoming state game warden, summered at White Grass from 1963 to 1965. He, like Karin, recalled no grizzlies on the ranch but agreed that black bears were commonly around, causing little fear among residents and guests and few "problems" at ranch or park dumps. Stories were told of bears trying to break into the walk-in cooler. Kitchen staff tried to run them off. One time in the 1980s, a bear was up a tree with about seventy ranch horses in the nearby corral.[11]

Cynthia Galey Peck first went to White Grass with her ranch-owner father in 1946, when she was three years old, and grew up there until she was sent to boarding school in her teens. The ranch had its own road, too long to be called a driveway. The dump was east of the "Main Cabin," which looks more like a lodge, just off what is now the Death Canyon road and near the ranch cemetery, in an area Cynthia described as full of natural sinkholes. Guests would drive to the dump to watch bears, which she said were around the ranch buildings all the time in the late 1940s and early 1950s. Since the ranch had guests for just two or three months each summer, workers fed some of the trash to the chickens and pigs. "Especially when only three of them [she and her parents] were there, there was very little trash . . . they'd burn it every fall . . . and of course, it decomposed." She remembered seeing one grizzly bear in that time. While wrangling her horse around a Douglas fir in Wister Draw, between White Grass and the JY, she had a surprise encounter when the bruin stood up close by and looked right at her.

She did not recall any bear-related human injuries in her time on the ranch, despite an instance in which some "really, really stupid cousins . . . one on horse, roped a cub and dragged it to the barn. After cutting the rope, the cub ran off and fortunately, they never saw mama bear." Cynthia, like Judith, mentioned that neighboring ranches—the Circle H, Bar BC, R Lazy S, and JY—all had dumps, but when they went visiting it was for picnics, dances, or barbecues, not to look for something as common as bears. She often took trips to Trail Creek Ranch or Sawmill Ponds (not far north and east, respectively, of White Grass) to see moose or elk. She remembered that some folks did go to the park dump at Timbered Island to see bears. By the late 1960s or early 1970s, "the park decided the ranch shouldn't use the dump and gave them four trash cans—not nearly enough," and workers were supposed to take them down to the gate for a park maintenance truck to pick up.[12] Exactly when White Grass Ranch ceased using its own waste pit is unclear. Trash dumping may have continued until the Galeys turned the ranch over to the park. At some point, private landowners were encouraged to take their trash to one of Grand Teton's sanitary landfills. An undated NPS permit was issued to Irwin Lesher, who until 1980 owned the X Quarter Circle X Ranch (north of Taggart Lake trailhead, where the historic Manges Cabin remains). The permit was for disposal of up to two large cans of household waste per day between June 1 and September 10.[13]

Like dude ranches, park concessioners generated trash during their operations and deposited it nearby in convenient locations. This practice, too, appears to have waned by the 1970s. When Grand Teton Park worked on its first Snake River Management Plan and considered limiting commercial rafting trips, staff proposed removal of an old garbage dump at the Deadman's Bar launch.[14] A previously mentioned 1982 report referred to a former dump at Flagg Ranch in the Rockefeller Parkway. Its exact location is unclear, but it likely was sizable to handle the extensive concessioner-run visitor services that were permitted in the area for years by the U.S. Forest Service.

For decades, the park itself dumped garbage at three major locations—south of the Colter Bay visitor services area; at the Kelly pit to the west of the small unincorporated community on the Gros Ventre River; and at Timbered Island, historically the biggest and most used of the landfills.[15] As its name implies, Timbered Island stands out in the sagebrush plain that banks the Snake River through the heart of Grand Teton National Park. It is actually a glacial moraine, about two and a half miles long and a half mile wide at its broadest point, marking a lateral edge of ice advance during the Bull Lake glaciation, which occurred between 150,000 and 45,000 years ago. The site of the dump can still be reached by a two-track dirt road on the south end of the moraine, west of the Snake River off what is known as the "Inside Park Road," a paved stretch closed to vehicles in the winter. When the Inside Road was unplowed, ranches likely held their refuse for months, as Cynthia Galey Peck recalled, while park crews dumped accessible trash at Kelly. And bears knew and made use of the park's dumps.

The archive's first official bear incident report comes from 1958, when the superintendent wrote to the regional director that "improved management of garbage dumps during the past two years—daily burning and covering of debris—has definitely increased bear activities at all campgrounds . . . and at various private ranches in the southern half of the park. Bear trapping has increased the past two years . . . Expeditious trapping of raiding bears has also apparently prevented development of roadside beggars, as none have been observed."[16] But, unsurprisingly, in following years this changed. Bears overturned garbage cans along the road near Timbered Island, near Jenny Lake, and in park residential areas. There were complaints from the Jackson Lake Lodge of a mother bear and cubs raiding garbage containers. Bear activity at dumps did not lessen—rangers estimated that ten bears occupied the area around the Colter Bay garbage site during the summer of 1959. From 1964 to 1968, bears regularly visited that dump and the one at Timbered Island, as well as those of private ranches along the Moose-Wilson Road. Park staff attributed three property disturbance incidents at the White Grass camp-site in 1967 to the concentration of bears at the nearby ranch dump. Despite landfills continually attracting bears and the associated reports of property damage, superintendents did not seem too concerned. For years, they wrote that "perversion of the food habits of the bear population" was the only "other" problem attributed to the dumps.[17]

Whether coincidental or not, in 1969, the same year that people reported rare grizzly tracks near Lizard Creek and an unconfirmed grizzly bear sight-ing on Hermitage Point, near Colter Bay on Jackson Lake, the superinten-dent atypically mentioned bear seekers: "The two Park dumps, Timbered Island and Colter Bay, continue to attract bears . . . Park visitors and local res-idents are also attracted to these areas nightly. Barricaded dump roads would eliminate the visitor problem and cessation of the land fill method of garbage disposal would eliminate the attraction for wild animals and birds. A high efficiency incinerator appears to be the best answer to the Park's garbage dis-posal problems."[18] Regrettably, the format of the park's annual bear activities reports changed in 1970 and 1971. Narratives were nearly eliminated or are missing, and subsequent reports fail to comment on whether any progress was made on those recommended actions; I found no evidence that the park ever invested in a trash incinerator. Nor is there further indication of how many people sought out Teton Park dumps to look for bears or for how long this dump visitation went on.

I was fortunate to talk with several former Grand Teton employees with personal knowledge of the landfills. Steve Baldock, a now-retired engineer-ing equipment operator supervisor and plow driver, grew up in Jackson Hole and started working for the park in 1991. He recalled picking up trash when the park had lots of garbage cans but very few bear-resistant dumpsters, and he shared other history he had experienced or had been told.[19] While the dumps were still open, at some point "an attempt was made to fence them, primarily to keep the trash from blowing around" rather than to keep out bears. He recalled seeing no grizzlies and believed black bears were fewer then than today. Baldock thought all three park dump sites were closed by 1970 or 1972, coincident with the closure of Yellowstone National Park's dumps, as is the recollection of another retired maintenance equipment operator, John McAvoy. He remembered a gate blocking the entrance to the Timbered Island dump, which "still looks like a hole" in the ground from which material was taken to build roads, and where oil tanks and more were stored during his tenure.[20] McAvoy said a big blowdown in the park sometime in the 1970s felled millions of trees, which staff spent years cutting and clearing from campgrounds before moving them to Timbered Island, where the old trash pit had been graded over.

Barry Alexander went to work in Grand Teton in 1970. His first job was to drive garbage trucks around the circuit between Timbered Island, Colter Bay, and Kelly. (The latter two maintenance areas are still in use as "dry dumps" to store items such as gravel or pipe, culvert, or other construction materials.) Alexander drove single-axle packers on a two-and-a-half-ton frame, which could fill up after he stopped at the Grand Teton Lodge Company—by then, he said, the park was hauling trash collected from the concessions operations as well as park residence and general visitor areas. Trash cans in the 1970s were thirty-five-gallon metal containers with plain flat covers chained on top, rather than "mailbox-type" lids or locks. Seasonal workers would ride in the back of the truck and place collected trash bags in the packer. He remembered all three dumps being fenced and said another employee would drive a dozer to each dump once a day to "tack things down in the wind." Once, after the sites were closed to dumping, someone dug up one of the trash sites to see what was there and commented, "Those hot dogs look brand new!"[21]

"I know there were bears around . . . not a lot of grizzlies then but lots of black bears" at all three dump locations, Alexander recalled, but he never had a close call or worried about the bears, even when dumping "consumables." "The bears didn't hang in too closely," he told me, but sometimes he would see "a black or brown butt running away in the trees" when he pulled his truck up to empty the trash. He said the park started hauling garbage to town in the 1970s and gated all the park dumps at about the same time. He replaced the gate at Colter Bay in 1979 or 1980 himself—though he said it was always left open because the man who ran the firewood concession split and stored wood there and needed to get in and out. (Apparently the park did not provide him a key.)

Documentation on precisely when and how Grand Teton National Park closed its in-house landfills is absent from local archives. However, coincident with the recollections of former maintenance men, references in Jackson Hole newspapers suggest the dumps closed in 1971 or soon thereafter, likely in reaction to several things. The U.S. Congress passed the first Solid Waste Disposal Act in 1965 and, with an environmental movement growing across the nation, established the Environmental Protection Agency (EPA) in 1970 to address concerns for clean air, water, and pollution. Subsequently, states began enacting their own regulations to address hazardous and nonhazardous solid waste, much of which had previously been dumped in open pits and burned. Parks would have been paying attention. Plus, news features and letters to the editor indicate heightened concern over Yellowstone's dumps being an attractive nuisance for bears, especially grizzlies. The public appetite for feeding wild animals, whether the average citizen watched or took part at all, had declined; biologists and conservation-minded individuals and organizations widely espoused a new ethic. It seems clear that the park ceased dumping bear-attracting solid waste years before it stopped using Timbered Island to store other unwanted and unsightly items. Now-retired biologist Steve Cain remembered the Park Service was still restoring that site when he reported to work at Grand Teton in 1989.[22]

At their meeting on August 3, 1971, Teton County commissioners discussed working with the city of Jackson, the U.S. Forest Service, and the NPS to survey how and where to establish a new site for trash, since the county site in South Park was soon to be closed. According to board chair Art Brown, "Whatever we decide, we know we won't be burning out of a pit again. We'll go to either the high incendiary or the fill method."[23] Several months later the town council and the county decided to retain the dump at its present location but change to a sanitary landfill operation.[24] Evidently the actual dump transition was not worthy of generating a news feature, but just over a year later, the *Jackson Hole Guide* reported that citizens in Teton County, Wyoming, could no longer burn or dump trash openly on private or public lands due to national and state air and water pollution laws; the county sanitary landfill (SLF) was the only public option remaining. "All other public open dump grounds are closed, including those in the National Park and U.S. Forest lands and on other federally controlled lands. Grand Teton National Park and the U.S. Forest Service will use the Teton County SLF for disposal of all its refuse and will pay for this service."[25]

A decade later, the county landfill was running out of space and, due to limited and pricey private property, managers seriously considered incinerating garbage to generate electricity, steam, or oil to offset costs and space associated with leaving trash on the landscape. Ironically, one of more than a dozen potential new landfill sites they identified was back in the park at Timbered Island.[26] Thankfully, that proposal was never implemented. The county capped and closed its landfill, hauling some 800,000 cubic yards of trash to other sites, and converted to a transfer station operation. As of 2023 Grand Teton and its concessioners trucked all garbage from their facilities, including the one remaining dude ranch in the park, the Triangle X, to the transfer station south of Jackson and paid to dispose of it. From there, Teton County sent trash on to Bonneville County, Idaho, a good distance from the bears of Jackson Hole.[27]

A "Moran News" local's column from October 1, 1970, reported a "Hot tip. A reliable source reported a grizzly bear nosing around the Colter Bay garbage dump! I don't think I'll go up that way."[28] From anecdotes like this, stories from dude ranch workers, and park records, it appears that while some people in Grand Teton did watch bears at dumps, the pastime was never promoted as it had been earlier in Yellowstone's history. Still, somewhere out there are photos that may turn up of black or even grizzly bears at one or more of the old park dumps.

8

Berries and More

I am a devoted hunter—of wild berries. Part of my regular summer routine for nearly three decades has been to search for one of our favorites, huckleberries, which my family enjoys eating in homemade pancakes, scones, jam, and pie, not to mention the occasional frozen daiquiri on warm summer evenings. I began "berrying" when my eldest daughter was born and I wanted to get outdoors but had only several-hour breaks between bouts of nursing my newborn. Holed up in our family cabin outside of Yellowstone National Park, I discovered huckleberries, finding them in more and more locations as years passed. With its predominantly volcanic soils and fairly dry climate, Yellowstone is not great berry country compared to wetter areas such as Glacier National Park and the Pacific Northwest, but I have found the best luck near its northeast and southwest corners. The Teton Range is locally well known for huckleberries, and when I moved to Jackson Hole, I quickly found favorite spots outside the park just a short drive from my home. Only once (so far—knock on wood) have I seen a bear in one of "my" berry patches. My youngest daughter and I noticed a large brown-colored black bear about seventy-five yards off the road as we pulled up to a productive area we had worked for hours the day before. Nervously she asked, "Mom, what should we do?" Safety and practicality made our next step obvious; I said, "We'll just find another place to go today," and we moved down the mountain to another location.

Picking berries, for me, has doubled as a sometimes-productive activity and as time to be calm and clear my mind, or conversely to mull over professional and personal challenges. Not uncommonly, as I keep alert, listening and periodically looking around for bears, I ponder whether I should leave the berries in the woods rather than provide a little competition, if meager, for the bruins. How important, you might wonder as I do, are berries to Grand Teton bears?

John Muir famously wrote of what he called the

> Sierra bear, brawn [*sic*] or gray . . . To him almost everything is food except granite . . . Every tree helps to feed him, every bush and herb, with fruits and flowers, leaves and bark; and all the animals he can catch . . . and ants, bees, wasps . . . A sheep or a wounded deer . . . should the meat be a month old, it still is welcomed with tremendous relish. After so gross a meal as this, perhaps the next will be strawberries and clover, or raspberries with mushrooms and nuts, or puckery acorns and chokecherries.[1]

Muir colorfully described bears as what biologists dryly term omnivores, eaters of both meat and plants. The list of foods for the now-extirpated California brown bears in the mountains of Yosemite at the end of the nineteenth century would apply as well to bears in today's Jackson Hole and surrounding areas.

A group of seasoned wildlife biologists scoured old and new accounts from 1891 to 2013 and, in a twenty-first century publication, noted what amounted to 266 foods (175 plants, 37 invertebrates, 34 mammals, 7 fungi, 7 birds, 4 fish, 1 amphibian, and 1 algae) plus one soil type known to have been consumed by grizzly bears in the Yellowstone ecosystem.[2] That study and much management discussion has focused on five major foods for the grizzly population—elk, bison, cutthroat trout, army cutworm moths, and whitebark pine.

In another study, comparing the diets of black and grizzly bears primarily in Grand Teton, researchers identified 134 different bear foods: insects (ants, wasps [Vespidae], unidentified types, and larva), fruits, nuts, vertebrates, and both above- and below-ground vegetation are named in the ecosystem-wide synthesis. Most frequently consumed above-ground plants were licorice-root (*Ligusticum* spp.), sticky geranium (*Geranium viscocissum*), cow-parsnip (*Heracleum maximum*), dandelion (*Taraxacum* spp.), brome grass (*Bromus* spp.), sedge (*Carex* spp.), lousewort (*Pedicularis* spp.), bluegrass (*Poa* spp.), fireweed (*Epilobium* spp.), horsetails (*Equisetum* spp.), clover (*Trifolium* spp.), and reed grass (*Calamagrostis* spp.). Below-ground plants included yampa (*Perideridia* spp.), onion grass (*Melica* spp.), biscuitroot (*Lomatium* spp.), spring beauty (*Claytonia* spp.), bistort (*Polygonum* spp.), and pond-weed (*Potamogeton* spp.) Bears ate the cambium—the growing part of a tree trunk that produces its bark—from five species, primarily lodgepole pine, the most common tree in the ecosystem. They consumed nuts from only one tree species, whitebark pine. (The area lacks elsewhere-common nut-producing types such as hickories, oaks, and walnuts.) Bears ate the fruit of twenty-one shrubs, most frequently huckleberries (*Vaccinium* spp.), buffaloberries (*Shepherdia canadensis*), serviceberries (*Amelanchier alnifolia*), gooseberries (*Ribes* spp.), mountain-ash berries (*Sorbus scopulina*), hawthorn berries (*Crataegus* spp.), chokecherries (*Prunus virginia*), and rose-hips (*Rosa* spp.). Unidentified mushrooms were lumped into the category of "fungi" as bear food.[3]

Note the absence of one of the more important grizzly bear foods, army cutworm or Miller moths, in park bears' diets. The small moths migrate to elevations high in the Absaroka Mountain Range in summer, where they estivate, burrowing into cool talus slopes in the heat of day. At night, the moths emerge to fly around and pollinate fields of alpine flowers. Dozens if not hundreds of grizzlies each year (surprisingly, as of 2023, no black bears) find moths an extremely valuable high-fat source of up to 20,000 calories per day at thirty-four known moth aggregation sites, all outside the ecosystem's national parks, although it is possible that a few grizzlies have home ranges large enough to include both those areas and portions of Grand Teton National Park.

Among the contributors to the latter study was Leslie Frattaroli, one of few scientists that focused on bears that live in and near Grand Teton

National Park. From 2004 to 2006, she researched black bears and their food habits.[4] As is typical across the continent, ants and other insects were a staple providing protein for black bears. Bears also ate graminoids (grasses), such as those listed above, especially in spring when few other foods were available. Grizzly bears are known to dig up and eat the fleshy underground root stems or corms of tasty forbs from the carrot family and other plant groups, but Frattaroli found no roots at her feed sites nor in black bear scats. While the bears' ease in digesting the high cellulose (fiber) content of cambium likely made it hard for her to find it in scats, she did see where they had stripped through outer tree bark with their claws to feed on the underlying layer.

One of the most valuable bear foods in the ecosystem is whitebark pine, whose large cones produce, every second or third year on average, bumper crops of nuts that appeal to bears, just as pinyon pine nuts appeal to certain humans. Grand Teton National Park and the Rockefeller Parkway have 28,500 acres of whitebark pine stands, and awareness of the food attractant was clear decades ago. A 1976 news article headlined "BEARS CLOSE CAMPGROUND EARLY" referred to Jenny Lake Campground, which was two-thirds full and then typically stayed open until the area received snow in late October. When multiple bears causing damage to campers' cars and property prompted the closure, the park's information officer attributed the increased bear activity at least partly to the year's large crop of whitebark pine nuts.[5] Three decades later, in Frattaroli's study, black bears made minimal use of whitebark pine seeds. In other areas the bears have been known, as are grizzlies, to find the large cones in red squirrel middens. Two of her nine collared study subjects climbed trees and broke branches to get at whitebark pine cones at eight sites in August and September, and she found seeds in scats from July into October. For those bears that can get them, whitebark pine seeds supply a high-fat source of protein, mostly above 8000 feet in elevation. The cones of limber pine, another five-needled conifer, provide food for black bears in some parts of the country, but although the trees occur in southern Grand Teton, no bears used them.

Park staff have monitored the effects of non-native blister rust, a disease-causing plant fungus, and native mountain pine beetle epidemics on whitebark stands for two or three decades. Mortality of sample trees has been high, and vegetation managers across the ecosystem have coordinated efforts to maintain whitebark pines, not only for their value as bear food, but for the species itself. Biologists have identified trees that appear to be naturally resistant to blister rust and send specially trained tree climbers to install temporary wire cages over the pine cones to prevent birds like Clark's nutcrackers from harvesting them. Later, workers collect seeds from those cones and contribute them to propagation efforts at Forest Service–led seedling farms. The seedlings can be used to plant whitebark pines in post-construction restoration areas or elsewhere, if needed. Although all whitebark pine transects checked in the ecosystem in 2021 showed signs of reproduction, the mountain pine beetle, a native insect whose boring activity weakens and can eventually kill trees, also occupied all the stands, and many trees had died.[6] Concern for whitebark pine across the Northern Rockies prompted the U.S. Fish and Wildlife Service to list it as a threatened species in December 2022.

As for the fruits of the forests, bears make much use of berries in summer and into autumn, especially huckleberries, of which multiple *Vaccinium* species occur in Grand Teton National Park. The southern park has, perhaps because of a slightly wetter climate and more granitic soils than the Yellowstone plateau, more visible available huckleberry patches, ranging from the shaded low portions of the valley to elevations near the tree line. I accompanied Frattaroli on one of her feed-site analyses, searching an area on the slopes above Phelps Lake where a radio-collared black bear had spent the previous few days. We found numerous scats, and she pointed out subtle evidence that the bear had been finding hucks. It was early in July, but we were surrounded by an abundance of already ripe purple berries that bears had missed or skipped during that feeding bout. I was struck by the bear's ability to delicately pluck the small morsels from the twigs; the animal had not, as I had assumed, stripped whole branches of stems, berries, and leaves. It reminded me that I once had occasion to watch a black bear moving slowly along a wooded trail, gently plucking wild strawberries from the ground-spreading plants, which afterward showed little sign of the bruin's grazing. Wild strawberries are even tinier than huckleberries; in the Teton ecosystem

FIG. 7. Black bear cub eating berries. National Park Service photo.

they top out at about the size of my fingernail. It appeared to me that both are worth a black bear's time and effort to obtain.

Next to huckleberries among shrub species, black bears eat serviceberries and hawthorn (*Crataegus douglasii*) berries. The latter are not widespread in the parks. Longtime Yellowstone Park botanist Jennifer Whipple knew of a single hawthorn plant near the Mammoth Terraces, and only one population of it in southwestern Yellowstone—practically in the Tetons. Dan Reinhart sampled grizzly bear feed sites for the Interagency Grizzly Bear Study Team from 1984 into the early 1990s then worked in Yellowstone until he finished his career in charge of vegetation management at Grand Teton Park; he found few hawthorns in either park except in moist, low-elevation areas such as along the Moose-Wilson roadway between the Death Canyon road and Sawmill Ponds.[7] Since at least 2010, both black and grizzly bears have often foraged in those bushes. Somewhat unfortunately, hawthorns also surround the Laurence S. Rockefeller, Jr. Preserve visitor center, designed before research identified the importance of that food to Teton bears. When the berries are "in," the dense shrubs make it easy for an unwary hiker to surprise a bear walking the trails from the center toward Phelps Lake, so park rangers often temporarily close the trails due to bears feeding there. Frattaroli recommended that planners of future developments consider proximity to hawthorn and other important bear foods.[8] In truth, most of the campgrounds and many trails in the park were built decades ago among high-quality berry patches that line the glacial lakes of the valley floor and canyons of the Teton Range. This sets up the continual potential for bear-human encounters in the most heavily used natural areas of the park.

Few black bear studies in the ecosystem have occurred where both species of bears live. But when Frattaroli did her field work in the early 2000s, black bears in the southern part of the park occupied an area *not* inhabited by grizzlies, making their population what biologists term "allopatric"—as compared to "sympatric," when the two species overlap in range, as was happening in northern Grand Teton National Park. All bears gained fat and lean body mass in summer and fall by including plant and animal food items in their diets, although southern Teton black bears made more use of berries—larger-bodied black bears and grizzlies cannot gain weight on a diet of berries and fruits alone. Researchers expected that black bears lacking direct competition from grizzlies might more often choose higher-energy protein such as ungulates, but they did not see it. In three instances, southerly black bears ate newborn deer fawns and elk calves in spring, though Leslie could not distinguish whether bears preyed upon or scavenged the young ungulates. In five cases, black bears fed on adult elk in the autumn. Nevertheless, black bears overall ate similar foods, regardless of whether they shared home ranges with grizzlies, sticking to a diet richer in ants and plants. Use of ungulates was, in scientific terms, "strongly associated with grizzly bears and greater body mass" as may also be seen in adult male black bears.[9]

Teton bears of both species have been known to eat at least eighteen vertebrate foods, from large and small mammals to birds and bird eggs—but they do not regularly find or eat fish. Cutthroat trout spawning streams, such as those that provided substantial food for bears around Yellowstone Lake in the latter decades of the twentieth century, historically may have provided spring sustenance for bears in Jackson Hole. Early in the 1900s, the addition and later augmented height of a dam on the natural Jackson Lake inundated the mouths of streams flowing into the lake, changing the water levels and substrate of spawning habitat and affecting the bears' ability to catch the park's only native trout. Also, the lake was invaded at about the same time by non-native lake trout washing downstream from Lewis Lake in southern Yellowstone, where they had been stocked by the former U.S. Fish Commission.[10] Just as the larger lake trout have competed with and eaten the much smaller cutthroats in Yellowstone Lake since being illegally planted there in the 1980s, so lake trout reduced, probably by 90 percent, native trout in Jackson Lake. Cutthroat trout still spawn in the lake's northern tributaries, some of which extend into the Rockefeller Parkway, but as of 2022 fisheries biologists had neither documented nor heard reliable reports of bears trout fishing in the park or parkway.[11] In one unusual instance, observers saw grizzly bear #399 and her two cubs fishing at Grand Teton Park's Oxbow Bend in the spring of 2008. The bears reportedly ate bluehead suckers, a native fish that spawns in the area. Some of the suckers and possibly other fish had become stranded in a pool, where they overwintered under the ice, presumably cut off from the Snake River when water releases from Jackson Lake Dam dropped in the fall. According to local wildlife watchers, coyotes had previously dug out the fish and buried remnants in the snow, where the grizzly family later found the cache.[12]

The most often consumed vertebrate bear foods in the valley were elk (*Cervus elaphus*) "by far," followed by deer (*Odocoileus* spp.), domestic cattle (*Bos taurus*), and moose (*Alces alces*). Bruce Smith studied elk on the National Elk Refuge, ultimately completing his PhD on the Jackson herd when about 35 percent of them summered in Grand Teton Park. In 1990–1992 he captured and tracked mortality of 164 elk calves, 126 of which were caught in their first week of life inside park boundaries and another 38 from the adjoining Bridger-Teton National Forest. Locations of calf deaths varied from their capture sites. Nineteen of the calves dropped their collars and could not be subsequently tracked, fifteen were killed by predators, and seven died of other causes.[13] Coyotes killed four calves, and black bears of unknown size and age at least ten, with another suspected; eight of the black bear kills occurred in the park. These predators killed calves between the ages of two and twenty-three days. Few grizzly bears were thought to live in Jackson Hole at the time, and Smith found no evidence of grizzly bear predation. Five years after his field work on that study ended, Smith again examined elk calf mortality in the eastern part of the park and adjacent forest lands, where cattle grazed. From 1997 to 1999 he radio collared 154 elk calves and documented 42 mortalities, 32 of which were caused by predators and 23 of those by bears. He identified with certainty six calves preyed upon by grizzlies but otherwise could not distinguish which bear species killed the elk.[14] Joel Berger also studied predator effects on large ungulates around the turn of the twenty-first century and found that grizzly bears had killed seven moose within the park.[15] The ursids also occasionally prey upon bison and often seek their winterkilled carcasses as a choice post-denning food, but little is known about bear use of bison in Grand Teton National Park. As of 2006 Costello and her collaborators had found only two instances, revealed in scat analyses, where one male and one female grizzly fed on bison in the month of June. Since grizzly bears and bison have both increased in number since then, this could have changed.

Bears occasionally kill each other. Some of these interactions seem to be driven by competition, such as when a male bear kills cubs, freeing a female to go into estrus and breed sooner than she could if she were still raising her young. This appeared to be the motivation for a spring 2022 interaction among grizzlies observed along the roadside in northern Yellowstone. Cases of outright predation also occur between bears. In late October 2014 an adult grizzly boar that had been trapped and radio collared in the Rockefeller Parkway covered the length of Grand Teton National Park in short order and headed for the Beaver Creek area where, unbeknownst to nearby human residents, he dug two adult female black bears out of their early winter dens in the period of a week. The grizzly stayed near his kills for several days each and consumed them.[16] Harsh though it may seem to humans, the black bears doubtless supplied a significant nutrient boost to the larger grizzly just prior to his own hibernation.

Bears benefit greatly from high-protein foods—meat, mostly—when they can get it. And one of the most noticeable sources of protein highlights a major difference between Yellowstone and Grand Teton, where the historical and current occurrence of livestock grazing has not gone unnoticed by the park's bears.

9

Bears and Livestock, Part 1

SHEEP

In 1997, when I was editor of a semi-technical journal published by Yellowstone National Park, the chief of resources and I interviewed Dr. Richard "Dick" Knight, the original head of the Interagency Grizzly Bear Study Team, prior to his retirement. I had worked with Dick for a number of years on the initial version of a conservation strategy for the ecosystem's grizzlies, and I had the pleasure of hearing him give semiannual reports to managers on the status of the population and his team's research. Those who knew Dick recognized he was quite a character. I liked the way he never became a "typical bureaucrat"—he was a well-trained independent scientist but looked like he'd be equally or even more at home on a farm or ranch. He was known for being very direct, even blunt, in responding to questions from forest supervisors and park superintendents, or from reporters and curious students of bears.

    Among things we discussed in the interview, I asked Dick about the effort to remove domestic sheep grazing allotments from what I think of as the "back side" of the Tetons—the Idaho side, where hundreds of sheep once ranged on the Caribou-Targhee National Forest. When I was in graduate school at the University of Montana in the early 1980s, a Forest Service Region One manager giving a guest lecture bemoaned that attempts to eliminate sheep allotments around Yellowstone had been unsuccessful. He was afraid it would mean the demise of the grizzly bear population. Knight said, "We were really involved in that. The Gallatin and Targhee [National Forest] supervisors just didn't want to believe that grizzly bears and sheep couldn't coexist. But we got the allotments out."

"So, sheep and grizzly bears are truly incompatible?" I asked.

"Well," Dick Knight replied, "it's the herders *that are really incompatible. Grizzly bears really* like *sheep!"*[1]

When people think of national parks, they seldom associate them with livestock. If they do, they might think of cows pastured in a historic eastern unit or of themselves taking a scenic ride by horse or even mule, as visitors do every day in Grand Canyon National Park. They might know that in the Tetons and other places, well before they became parks, American Indians, trappers, and miners used stock to cross the western mountains, to explore, and to hunt. Today, in Yellowstone and many other iconic park units, visitors and staff need not worry about livestock conflicts with wildlife, because stock grazing does not exist save for the occasional string of park horses and mules used for trail work or mounted patrol—visitors love to photograph uniformed rangers sitting tall under their flat hats atop well-trained equines.

Actually, at least ninety-four units of the National Park System permit goats, sheep, even cows to graze within their borders, notably including Grand Teton National Park.[2] Not long before its major expansion in 1950, the park had eleven of its own stock, along with thirty owned by concessioners and seventy-five by private users; guest ranches in the park such as the Bar BC, White Grass, and JY grazed horses both on and off federal land.[3] Although there were occasional complaints of bears making the stock nervous, park records reveal no major conflicts, nor loss of horses or mules to bears. Some of the ranches had other farm animals. In 1962 the caretaker at the 4 Lazy F Ranch, just upriver from Moose, reported that his dog had killed a yearling bear cub, reportedly weak and starving, just out of hibernation, as it tried to kill a pig in its pen.[4] Cynthia Galey Peck recalled that at White Grass Ranch her father, Frank Galey, once shot at a black bear that had killed a pig and buried it in "one real carcass hump . . . [with] two or three false humps of twigs and logs" nearby in an apparent effort to conceal the true carcass location. The bear was a female in poor condition, with two cubs, and "Dad shot over her head and chased her away." She did not remember any other bear-livestock encounters at the family's ranch.[5]

Early Jackson Hole settlers tried to farm and raise livestock in the 1920s and 1930s in the southern part of the valley at South Park, along the Snake River south of the JY Ranch, and from Mormon Row and Blacktail Butte (areas not far from today's park headquarters, added to the park in 1950; see map 4) north to the edge of Sargent's Bay on Jackson Lake.[6] From the onset, there was a bias toward cattle over domestic sheep. In 1901, when issuing its first livestock grazing permits, the Teton National Forest prohibited sheep grazing to leave more forage for cattle and horses; a petition circulated in Jackson Hole argued that sheep would destroy the range.[7] Domestic sheep grazing was never allowed in the park. As early as 1919, when Wyoming's elected U.S. representative Frank W. Mondell addressed support for proposed expansion of Yellowstone National Park southward into Jackson Hole, he emphasized that "if . . . [it] would have the effect of reducing grazing privileges and opportunities in the state, either for sheep or cattle, I should certainly not be in favor of the extension . . . [but] not a single head of sheep is grazed within the area proposed to be added to the park. Not a single permit to graze sheep is, or has been issued."[8]

Sheep ranching was always more common on the west side of the Tetons, where Idaho sheepmen had opposed proposals that would expand Yellowstone, due to "the attitude of national parks *in general* regarding sheep, and the fact that livestock grazing privileges had been curtailed."[9] Some trailing of domestic sheep across the Teton crest did occur. In September 1929, when new Grand Teton Park superintendent Sam T. Woodring encountered herders with 1,742 sheep moving eastward over Fox Creek Pass and across the Death Canyon Shelf to the head of Teton Creek, he granted them permission on the spot since the park legislation grandfathered in existing livestock drifts and grazing.[10] In 1945 Superintendent Paul Franke honored a U.S. Forest Service permit that had existed since before the park's establishment allowing some 1,500 head of sheep a two-day crossing twice a year at the head

FIG. 8. Sheepherder with his flock (Madison County, Montana), 1939. Library of Congress, Prints & Photographs Division, *Look* magazine photo collection, photo by Arthur Rothstein, LC-USF33–003081-M2.

of Death Canyon, between Targhee National Forest grazing allotments.[11] In 1950 four Jackson Hole ranchers owned just 815 sheep, and by 1954 the number had dwindled to 104.[12]

Due to the small number of domestic sheep in or near the park and the mid-twentieth-century absence of grizzlies, there are few records of interactions between the two species. However, the 1975 listing of grizzly bears as threatened focused closer attention on the potential for conflicts, particularly on national forest lands that had long offered livestock grazing opportunities for local ranchers. Sheep provide considerable temptation to carnivores. And for much of the twentieth century, the stock found summer range in the grassy meadows off the Grassy Lake Road and among the Douglas fir, lodgepole pine, and aspens on the western foothills of the Tetons upward of 9,000 feet in elevation.

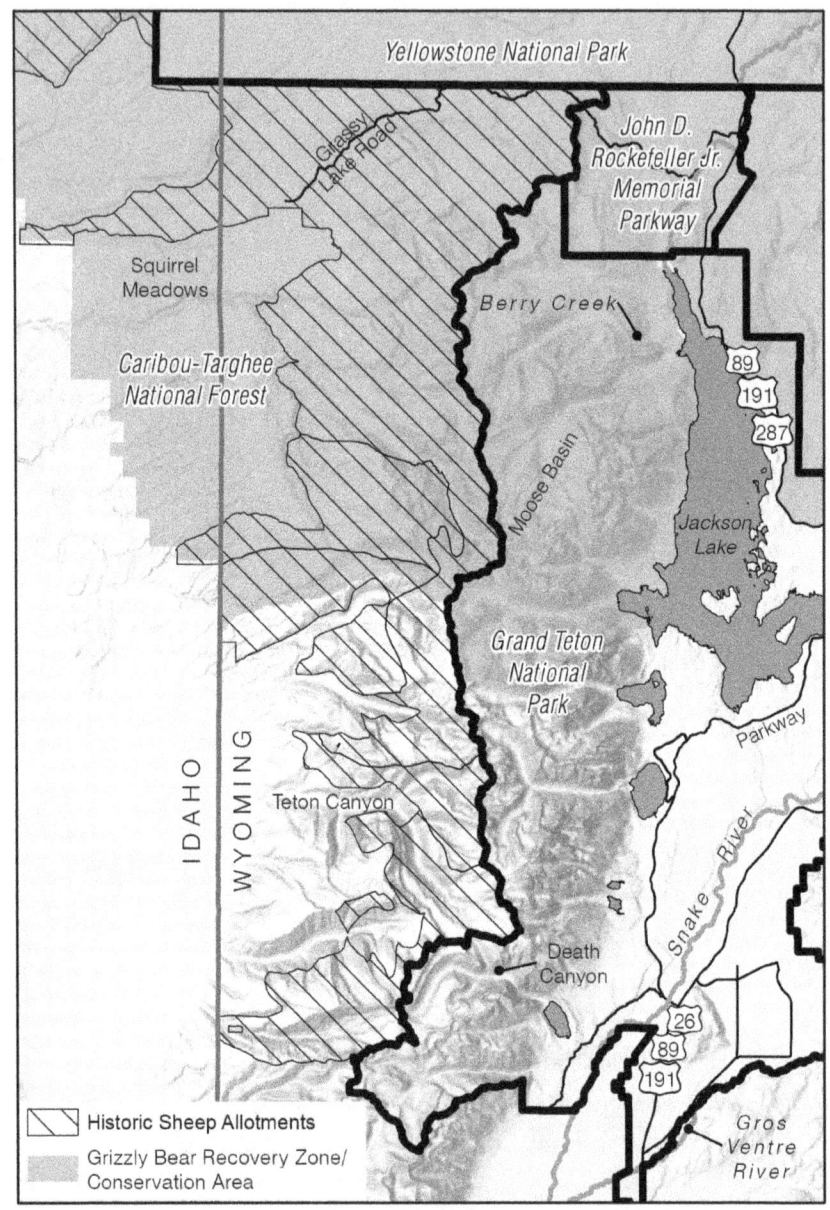

MAP 3. Historic sheep allotments near Grand Teton National Park and Grizzly Bear Recovery Zone. Map by Megan A. Smith, EcoConnect Consulting, Jackson WY.

Two Forest Service employees monitored grizzly bear predation on allotments for 15,707 sheep that grazed two areas near Yellowstone Park, one beyond the border town of West Yellowstone, Montana, and the other around Squirrel Meadows, Wyoming, where two of the studied allotments were on that park's southern border.[13] As is common even today, there was no mention of the latter's proximity to the recently designated Rockefeller Parkway less than fifteen miles to the east, nor to Grand Teton National Park, which bordered two of the Targhee National Forest grazing allotments northwest of Jackson Lake. And although I emphasize that the permitted sheep and the associated research were outside national park boundaries, their proximity to and influence upon bears make it worthwhile to include in a history of wildlife in the Tetons.

While it is hard to know how many bear-sheep conflicts occurred prior to the 1970s, the forest's study revealed sheep losses to both black and grizzly bears, as well as to coyote predation, disease, poisonous plants, and what researchers labeled "poor herding techniques." Black bears killed 196 sheep in 1976–1977 and, as a result, an associated thirty-one black bears were killed. Another eighty-four sheep were reported killed by grizzly bears, six of whom were "identified as sheep killers," but since grizzlies were by then protected under the Endangered Species Act, none were removed at that time. The study's authors discussed implications only for sheep-grizzly conflicts, not mentioning black bears in their summary, even though black bears killed more than twice as many sheep.

Carole Jorgensen conducted her master's degree research in the same area, coincident with the forest's monitoring effort, capturing and tagging black bears while the Interagency Grizzly Bear Study Team caught and marked grizzlies. Bears typically swatted down and killed sheep, often dragging them into secluded areas to feed on them. Sheep losses on three allotments adjacent to the park or parkway ranged from 25 in 1975 to 191 in 1977. Despite what the researcher described as challenges in differentiating between bear kills and those of coyotes, when little remained of most sheep carcasses, herders and federal agents alike attributed predation to bears based on their tracks, scat, and sign in the area, which she wrote "led to presumptions of guilt." In 1976 alone, she documented seventeen bears of both species that were killed as alleged depredators, ten by herders and permittees and seven by U.S. Fish and Wildlife Service predator control agents. Another eighteen bears were killed in 1977, fifteen by herders. There was negligible compliance with requirements for salvage of bear carcasses, and Jorgensen concluded that few of the bears killed by herders or permittees in response to real or alleged predation were reported. She suggested that, at least for black bears, true mortality in 1976–1977 was perhaps more than twice what was reported. To her credit, she did acknowledge the adjacent parks and their relationship to sheep grazing at a time when most people believed grizzlies were still absent from the Tetons: "Allotments consistently having the greatest bear depredation problem . . . all border[ered] either Yellowstone National Park or Grand Teton National Park." She speculated that those ungrazed units might serve as temporary sanctuary for bears whose home ranges overlapped park boundaries.[14]

The northwestern quadrant of Grand Teton Park then and now receives light human visitation, especially when compared to popular backcountry hiking routes such as the Teton Crest and the Cascade and Granite Canyon corridors. Few park staff or visitors venture into the less-accessible high-country trails of Moose Basin, or the canyons of Owl, Webb, and Berry Creeks just over the Tetons from the historic sheep grazing allotments—country that felt to me, in my twenty-first-century sojourns, wilder than other parts of the park. People believe what they see, and bears there have been "out of sight, out of mind" for most. However, as Rocky Barker, former longtime reporter for the *Idaho Falls Post Register*, wrote in *Saving All the Parts*, the west slope of the Tetons "had become infamous among grizzly researchers and managers as a place where grizzlies entered but rarely left."[15] In its earliest years, between 1974 and 1979, the Grizzly Bear Study Team trapped and followed thirty-seven radio-instrumented bears; seven of ten livestock-killing bears killed sheep. In response, two grizzlies were reported to have been killed on sheep allotments, but based on information from undercover agents and sheepherders, at least three and up to fourteen more grizzlies might have been killed in a two-year period. Team leader Dick

Knight, with his customary bluntness, wrote in his annual report (using a line omitted from a later-published conference proceedings article), "A good undercover agent with a 6-pack can probably get more information on bear mortalities in a couple of hours than more conventional methods could produce in 5 years."[16]

Barker noted that rather than stop the grazing, the Forest Service sent biologists into the field to keep an eye on the sheepherders and, while it immediately improved bear survival, the "uneasy peace didn't last long." Longtime U.S. Forest Service employee Dan Tyers, who eventually became that agency's grizzly bear habitat coordinator for the Yellowstone ecosystem, was originally hired in 1978 to, as he put it, chase sheepmen in the backcountry of the Gallatin National Forest. He presumed that someone was doing the same job for the Targhee Forest. During his years on sheep patrol, Dan would seek out herders in their high-country camps while checking for compliance with new grizzly bear recovery measures. One of those sheepmen, after a lifetime of summers tending flocks long before the Endangered Species Act existed, viewed grizzlies as sheep killers, a typical attitude of herders. Over time, Dan and the man developed a regular banter; after riding into the permittee's backcountry sheep camp, the Forest Service employee would get around to asking, "Have you killed any bears today?" And the old herder would respond with a smile, "Not today, Dan, not today," accounting for neither the previous nor a future day. With that predictable greeting out of the way, their conversation could continue. For Dan, that exchange came to represent the challenge of moving from an established way of ranching life into a new era working toward grizzly bear recovery.[17]

It was neither a fast nor an easy transition. Knight's study team was not subtle in raising concern for mortality of the recently listed bears, concluding that sheep and grizzlies were not compatible.[18] Researchers noted the existence of organized poaching rings north and west of Yellowstone and along the west side of Grand Teton National Park and the adjacent Teton Basin. Sheepherders openly admitted to getting rid of any bear they saw, whether it was killing sheep or not.[19] By 1981 the study team urged that, to recover the ecosystem's grizzly bear population, every reasonable effort must be made to reduce the animals' deaths to zero, listing garbage dumps and livestock allotments as the top two known controllable sources of mortality.[20] Their recommendation to eliminate sheep grazing within grizzly bear range was reiterated in a National Park Service Advisory Board report to then–secretary of the interior James Watt. And they specified the grazing allotments in the Targhee National Forest and their proximity to Yellowstone without mentioning Grand Teton National Park.[21]

Concern in many quarters for dead grizzly bears was understandable. But grazing had occurred in eastern Idaho's Teton Basin for nearly a century on both private and public land. And while not all bears kill sheep, bears of both sexes and varied age groups (subadults, adults, females with and without cubs) did. Predation is neither pretty nor, in some cases, limited to what an animal might need to survive. In the Squirrel Meadows area, two young grizzlies killed or injured thirty sheep in one night.[22] Bear conflicts, combined with losses from other sources such as coyotes, mountain lions, and disease, heightened the challenge of raising domestic sheep. The family of stockman Jim Davis had grazed sheep on the west slope of the Tetons since 1944. But by 1983 that area (along with nearly all of Yellowstone, part of Grand Teton National Park, and much of the wilderness in the ecosystem) had been designed as habitat where grizzly bears were to be given priority if there were conflicts with humans. In 1983 Dr. Chris Servheen, the first designated Grizzly Bear Recovery Coordinator for the U.S. Fish and Wildlife Service and the person responsible for oversight of the bears' recovery plan, said that the Forest Service would end sheep grazing in vital grizzly habitat within five years.[23] In 1987–1988, after a bear got into the Davis's sheep, forest representatives met with the rancher to discuss moving his stock to different allotments. Davis said the agency had to provide a place equal to or better than what his sheep had been using, but since the "West Slope is the best sheep range . . . left in this country . . . we can't win."[24]

Although Servheen underestimated how long it would take to phase out or relocate sheep allotments, both federal agency staff and ranchers kept working on it. In 1996 a female grizzly and two cubs suspected of killing ten sheep were moved from upper Badger Creek, west of Mt. Moran and the ridge of the Tetons. Evidence suggested at least one other grizzly was in the same allotment a few days later.[25] As of 1998, when the greater Yellowstone grizzly bear population was first within the biological recovery parameters, eighty-eight livestock grazing allotments remained in the recovery zone; eleven of those were active sheep allotments and another four were vacant of sheep.[26] Six years later, the Targhee National Forest documented having closed sixteen allotments to sheep grazing, including all but one of those adjoining Grand Teton National Park.[27] To assist a graduate student project on livestock grazing in greater Yellowstone, Dan Tyers visited all the Forest Service districts over a two-year period in the 2020s and looked at records of grazing allotments, the number of AUMs—animal unit months—permitted for sheep and cattle, and when allotments closed. He reaffirmed that sheep allotments were gone from the grizzly bear recovery zone by about 2004.

Even taking more than two decades of effort, the cessation of sheep grazing in optimal grizzly habitat was a significant conservation accomplishment by Forest Service officials—and a major change for individual ranchers who altered their working locations if not their traditional means of livelihood. It also had an unknown but, in this author's opinion, undoubtedly positive impact on the occupancy and expansion of grizzly bears southward into the Teton Range and into Grand Teton National Park. Somewhat ironically, and perhaps inevitably, greater Yellowstone grizzly bears have substantially expanded their occupied habitat in the past three decades, overlapping additional livestock grazing allotments outside the recovery zone. Occasional conflicts still cause both sheep and bear mortality in areas more distant from the national park units. The late Dick Knight's words still remind us that grizzly bears and sheep are not made to peacefully coexist. Thanks to his and many others' efforts, by the time I went to work in Grand Teton, domestic sheep were gone from the Teton crest. But there is more to the livestock-bear story.

Bears and Livestock, Part 2

CATTLE

When I worked in Badlands National Park, South Dakota, and even in Yellowstone, a cow would occasionally be reported in the park. They were not allowed to graze in either, but now and then one wandered across the border from a neighboring ranch in the Badlands, or from a grazing allotment on a national forest adjacent to Yellowstone. A patrol ranger was typically dispatched to try to herd the cow back to where it belonged. At Grand Teton, however, there are places and circumstances where cattle are allowed to be within the park. One day, at the historic Elk Ranch, I bumped along on a government-approved off-highway vehicle, or "OHV"—equipped with seat belts and roll bar—with the seasonal employee hired to mend stock fences and irrigate the fields. He was an amusing Texan character with decades of experience on ranches before coming to work for the park. He stopped periodically to show me his rudimentary system of flood irrigation using rocks to hold down tarps, making small check dams in the ditches that filled then spilled water over the surrounding grasses. As we worked our way across the pastures, he talked of intermittently seeing an early-morning wolf walk by, and of chatting with the researchers when they came to set bear traps nearby. He'd had no trouble with either predator bothering the cows but was ever vigilant for the possibility. I don't recall whether I told him that, prior to working in the Tetons, I first learned that cattle grazed in the park due to a newsworthy incident involving several grizzly bears killing cows in these fields . . . an incident that, more than a decade later, the superintendent and I were determined to make sure was not repeated. I also recalled hearing Wyoming bear biologists talk about trapping, for a special study, some of the largest grizzlies in years, weighing well over 600 pounds. On a 1993 late autumn hike out of Yellowstone's Thorofare area, my companion and I followed six-and-a-half-inch-wide, foot-long grizzly tracks for about three miles up and over the dusty Gravel Mountain trail to the Pacific Creek Road in Jackson Hole, and I thought, "That's likely one of the cattle-killing bears of the Tetons. I'm glad he's ahead of us."

Cattle have been a traditional part of the Jackson Hole landscape since Euro-American settlement. And they have continued to graze in the twenty-first century to the general surprise of visitors who, if they notice cows just off highway 89/287 near the east boundary, may assume that the cattle are outside Grand Teton National Park. To the contrary, the Elk Ranch, just south of Moran Junction, is the major remaining permitted livestock grazing pasture inside the park.

It was, in the early twentieth century, the largest cattle ranch in Jackson Hole. In 1914 Josiah Davis "Si" Ferrin secured a contract to supply beef to Reclamation Service crews building the Jackson Lake Dam. By 1920 Ferrin, the so-called Cattle King of Wyoming, had up to two thousand cattle on the Elk Ranch. When livestock prices fell a few years later, he sold his ranch to John D. Rockefeller Jr.'s Snake River Land Company.[1] The severity of Wyoming winters deterred most settlers from finding a stable livelihood in traditional cattle ranching, and most turned to dude ranching in addition to or instead of raising cows.

The 1950 legislation provided inholders—owners of private land parcels inside park boundaries—continued rights to graze nearby pastures unless their lands should someday be vested in the United States.[2] Dozens of small parcels not associated with livestock, such as home sites in Kelly and on Mormon Row, were incorporated into the expanded park, and some are still pri-

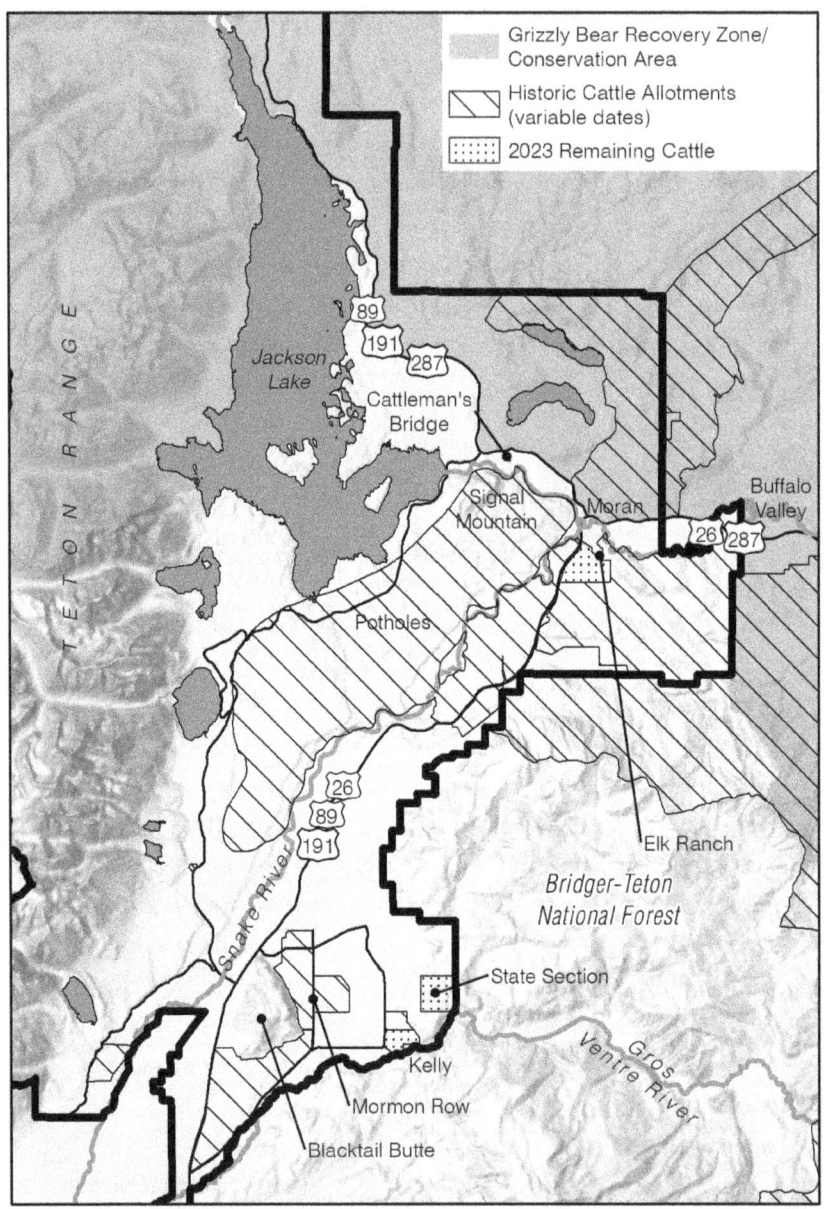

MAP 4. Historic and remaining cattle allotments in Grand Teton National Park. Map by Megan A. Smith, EcoConnect Consulting, Jackson WY.

vately owned. But just five ranches in the park grazed stock as of 1950, and only two remain in the 2020s—the Moosehead Ranch near Spread Creek, which grazes only horses, and the Pinto Ranch, spanning the eastern park border along the Buffalo Fork River. Though even longtime valley locals often think the Triangle X Ranch is a private inholding, its second owner, John S. Turner, was one of the first to sell his ranch to John D. Rockefeller Jr., in 1929, and he retained a lease so that his family could continue its guest ranch operation. Since 1954 the Triangle X has operated under some type of concessions contract authorization with the National Park Service (NPS), which also governs how they manage livestock grazing.[3] For more than half a century, the Triangle X Ranch has kept only horses needed for their business, serving summer guests and providing outfitter hunting trips outside park boundaries. But their original lease agreement allowed the Turner family to raise cattle, seventy-five of which were permitted to graze what was first Forest Service land and later transferred to the expanded park.[4] According to Harold, one of three Turner sons who ran the ranch for decades:

> Everybody had to have cows, because it's the only thing you could borrow money on . . . Horses weren't worth anything; land was worthless . . . we had cattle until 1965, when the Park . . . took away our cattle grazing permit [according to historian Robert Righter it was officially cancelled in 1967] . . . I thought it was the end of the world . . . when I look back on it, their taking our permit was probably the best thing that ever happened to us, because we put a tremendous amount of resources into managing the cows . . . there's a lot more money in the guest business in this valley.[5]

From the early 1900s through the 1940s, other ranches in the valley annually moved cattle from their own pastures, nearer the town of Jackson, to summer range on private or U.S. Forest Service lands in the heart of what became Grand Teton National Park. When tracts acquired by the Snake River Land Company, including the Elk Ranch, were donated and combined

FIG. 9. Cowboy with cattle in Jackson Hole, Teton Range backdrop. Jackson Hole Historical Society image 4191.01.

with additional Teton National Forest lands to increase the size of the park, the potential cessation of grazing by twenty-six ranches that had summer cattle leases was one of many contentious issues addressed in legislation establishing the new boundaries. The external ranch owners included some of the valley's most well-known early settlers, such as Robert Bruce Porter, who established the Jackson Hole Hereford Ranch and whose descendants in the Gill family still operate it; the Moulton family, original homesteaders on Mormon Row; and the Mead-Hansen family, which has provided two of Wyoming's governors—Clifford Hansen, who led the state from 1963 to 1967 and was subsequently a U.S. senator, and his grandson Matt Mead, who served from 2011 to 2019. The landowners were granted continued park grazing privileges for their lifetimes and those of one generation of their heirs, but that would cease sooner if their home "base" lands were sold or converted to nonagricultural use, or if they failed to exercise grazing for several consecutive years. A unique case is the Teton Valley Ranch, located along the Gros Ventre River and the park's southeastern border at the unincorporated community of Kelly, Wyoming. In the late 1940s, when he still owned the adjacent land, Rockefeller granted its owners, the brothers Wilson, a lifetime grazing lease.

In 1950 nongovernment users ran cattle on between 22 and 27 percent of Grand Teton National Park.[6] The four largest permittees formed the Potholes-Moran Grazing Association and grazed 1,154 cows on 83,500 acres each summer in a large open expanse of sagebrush grassland southeast of Signal Mountain.[7] The park built a "Cattleman's Bridge" so ranchers could drive their cows south across Snake River to the Potholes, a natural depression in the glacial outwash plain. Although used for only a few years, the simple log bridge remained until 2001, when it collapsed and was removed. The site can be found today at the end of an unpaved side road between Jackson Lake Junction and the Oxbow Bend pullouts. In 2010 the Wyoming State Historic Preservation Office provided a large sign interpreting the historical access.

In the enlarged park's first decade, cattle roamed well beyond the Potholes. Former rangers Maynard Barrows and Russ Dickinson (who later became NPS director) recalled cattle bedding down under the porte cochere at Jackson Lake Lodge and running through Jenny Lake Campground and the String Lake, Pacific Creek, and Colter Bay areas.[8] Park staff questioned whether cows competed with native grazers for forage and, in the latter half of 1951, reported that cattle had severely overgrazed the Hermitage Point area. They also photographed damage to alpine plants and scenery at Lake Solitude caused by a single party of ninety horses and mules.[9] Archival records tell of perennial discussion among biologists and some constituents about whether park pastures should serve cattle rather than the Jackson elk herd, but no apparent Park Service concern for bear-cattle conflicts.

By about 1957 the park began to move cattle grazing permits from the Potholes to the Elk Ranch and adjacent fenced areas beyond the east banks of the Snake, consolidating livestock into smaller areas so visitors might experience more native wildlife. Managers promoted better grazing at the pastures, which were, and still are, irrigated using water diverted from Spread Creek Dam into Uhl Hill Reservoir, toward the east end of the old ranch near the park boundary, or directly into ditches running from the dam (which was removed in 2010 and replaced with a diversion that permits upstream fish passage to the major Snake River tributary).[10] By the early 1970s only six of the original ranches outside the park continued to graze cows in those pastures or areas around Mormon Row that were used in early summer.[11]

Until the 1990s park records make no reference to damage caused by bears to cattle or cattle operations. Although at least two male and two female grizzlies elsewhere in the ecosystem killed cows between 1974 and 1979, and in two cases bears killed horses, none of the livestock killed were on the Targhee or Bridger-Teton National Forests, nor in Grand Teton National Park. Bear biologists until then had concluded that cattle allotments did not pose a serious threat to the grizzly bear population. It seemed that cattlemen, unlike sheepmen, did not kill depredating bears.[12]

But the big carnivores had killed cattle in the area before. A front-page headline in an August 1933 *Jackson's Hole Courier* declared "Two Grizzlies Guilty of Recent Cattle Raids" near Togwotee Pass.[13] A decade and a half later, renowned biologist Adolph Murie and his wife, Louise, along with his brother and sister-in-law, Olaus and Margaret "Mardy" Murie, lived at what is now the historic Murie Ranch at Moose. Though more famous for studying wolves in Denali (then called Mt. McKinley National Park) and coyotes in Yellowstone, Adolph published a 1948 paper describing "Cattle on Grizzly Bear Range" in Jackson Hole.[14] It was the first scientific publication documenting predation by bears on cattle in what is now Grand Teton National Park and in the adjacent Bridger-Teton National Forest. Cattle grazing allotments then ranged from low sagebrush and riparian corridors, along the confluence of the Buffalo Fork with the Snake River near Moran Junction, to higher-elevation meadows along Flagstaff, Grizzly, and other creeks flowing from Togwotee Pass, about twenty miles east of Jackson Hole.

Both black and grizzly bears were present from late June into August in 1945 and 1946, when about 1,650 cattle grazed the 88,000-acre Blackrock–

Spread Creek allotment just east of Moran Junction. Bears killed at least ten cattle the first year of Murie's study and seven the following year. In at least three more cases, bears tried but failed to kill young cows. An additional 1,200 cattle grazed three adjacent allotments at Pacific Creek, Lava Creek, and Moran and, although he did not actively study predation there, Murie noted at least two more reported bear kills. He described in detail each kill he inspected: Predation occurred not only at night but even around noon on summer days. Ravens flew or perched overhead in trees, occasionally holding a piece of meat. Tooth wounds were common on the necks and shoulders or backs of the kills, sometimes but not always crushing the vertebrae. He also noted how cows' livers, lungs, and hearts were left in place in some instances but eaten in others and how claw marks indicated a bear had turned over or dragged its meat. Murie's findings contradicted the traditional thought that bears killed with a crushing swipe of forepaw to skull. Instead, the predators used their front legs like arms to seize a cow then inflict a killing bite. He speculated that grizzlies did the killing, not black bears, but did not have proof in all cases. Livestock owners across Wyoming at that time considered black bears a "serious menace"—government trappers killed at least eighty-two just in 1941 and 1942 in addition to an unknown number killed legally by stockmen who believed they were a threat.[15] However, there is no record to indicate that black bears were a major predator of cows in Jackson Hole. On the other hand, both male and female grizzlies took cattle, all calves or yearlings.

Cowboys urged drastic control of all bears, even small black bears, and responded by sometimes trapping and moving a bear. If the predator returned to its prey, the wranglers might track the bear with hunting dogs and kill the suspected offender. Murie weighed the pros and cons of selective predator control and concluded that, while there were conflicts between grizzlies and cattle on what was then a Teton National Forest allotment, stock losses were relatively light. The grizzly, he wrote, being restricted in range, should be given special consideration. There is no sign of any agency follow-up to his recommendation. Instead, grizzlies declined or disappeared in much of Jackson Hole between the 1940s and the 1980s, and concerns for potential bear-cattle conflict went quiet, while grazing went on.

As of 1991 about 1,218 cattle belonging to the Mead-Hansen, Moulton, and Porter-Gill families continued to graze the Spread Creek, Cunningham, Elk Ranch, and Uhl pastures, south of the Buffalo Fork River extending eastward from the outer park highway to the national forest boundary. Another 150 Pinto Ranch cattle were permitted north of Moran Junction on the Pacific Creek allotment, of which about one-third was in the park with the rest in the national forest.[16] Paul Walton ranched in Spring Gulch and had once grazed cattle in Grand Teton Park, although he had lost that privilege in 1975.[17] But he retained a trailing permit to drive 750 cow-calf pairs, forty bulls, and nearly thirty yearlings across the Gros Ventre River, past Blacktail Butte to the Elk Ranch, and on to the large Blackrock–Spread Creek grazing allotment between the park and Togwotee Pass on the Bridger-Teton National Forest. This cattle drive reportedly occurred at a rather slow pace.[18] At meetings in 1991 through 1993 with park staff, grazing permittees conveyed their concerns: Who had responsibility to maintain pasture fences? Could park bison, potential carriers of brucellosis, transmit the disease to cattle? Could their presence force ranchers into more expensive testing or management of cattle to meet related state and federal requirements? There was of yet no mention of concern about wolves, despite public meetings and hearings being held in those years over their proposed restoration to the wilds of both nearby Yellowstone and central Idaho. Nor was there any mention of bears.

Then, half a century after Murie's study, the big predator made its presence known in another headline-generating way. As Andy Russell wrote decades ago, "when, through constant exposure to cattle, certain grizzlies acquire a technique for killing and a taste for beef, they can develop into avid predators."[19] Paul Walton had run cattle in the area since about 1959 and, according to his range rider Terry Schramm, reasonably coexisted for a number of years with grizzlies.[20] But up to nine grizzlies were thought to roam their summer grazing allotment, and the Wyoming Game and Fish Department confirmed that some of them killed six calves in 1992, then twenty-five more in 1993.[21] Mark Bruscino, now retired, had been hired in 1991 as the state's predator conflicts coordinator for northwest Wyoming. He recalled that livestock depredations by bears started on ranchland just east of Togwotee Pass in the Dunoir country, in the late 1980s and early 1990s, then moved to the Blackrock allotment.[22] It was a new challenge for field biologists, who had not experienced bears preying on cattle when grizzlies roamed mostly inside Yellowstone National Park or on ungrazed wilderness lands, and for senior decision-makers interpreting policies for managing the threatened species.

The Pacific Creek grazing pasture north of highway 287, the road from Moran to Dubois, and other grazed national forest lands east of the park boundary were in the grizzly bear recovery zone, split between Management Situation 1, where protection of grizzly bears is primary, and MS 2, where nuisance grizzlies were to be controlled in favor of other uses such as grazing.[23] In response to cattle killed in 1993, state biologists caught one male grizzly, #209, on a Grand Teton National Park allotment and relocated him to central Yellowstone, where there were no cows to tempt the predator. That fall, Wyoming Game and Fish officials and the Grizzly Bear Recovery coordinator asked Yellowstone Ecosystem Subcommittee managers to consider relocating adult male grizzlies suspected of killing cattle, though environmental groups criticized the proposal.[24] Ultimately managers stuck with their policy, from the Interagency Grizzly Bear Guidelines, of not moving bears from lands designated as MS 1.

The increase in cattle losses did prompt a three-year study by state game managers, starting in 1994. Mark Haroldson with the Grizzly Bear Study Team recalled then–team leader Dick Knight encouraging the Wyoming Game and Fish Department to step up their active participation in the team's research program, and this was a prime opportunity.[25] Bruscino pushed to better understand just how many bears were responsible for depredation. Chuck Anderson, who had recently earned his master's degree studying moose, was put in charge of the study along with Dave Moody, who went on to lead the state's large-carnivore research group, headquartered in Lander, Wyoming. Bruscino, because of his bear trapping experience, initially led both research and management captures associated with the cattle killings. The 110,700-acre Blackrock–Spread Creek study area was mostly adjacent to Grand Teton National Park but included the park's Elk Ranch East allotment, where about

900 cow-calf pairs grazed from early July to late October.[26] The Grizzly Bear Recovery Zone includes Bridger-Teton Forest lands south of Moran Junction, but not those pastures in Grand Teton National Park. Thus, managers were legally free to determine whether and how to manage cattle-killing bears on the Elk Ranch.

State biologists captured and monitored seven radio-collared grizzlies throughout the study area that first year and confirmed thirteen cattle killed by bears. Although twenty-nine cattle died of natural causes, another nine calves died from undetermined causes. Just two bears were responsible for about 70 percent of the calf kills.[27] Two of the cows and two calves died in Grand Teton National Park, where staff concurred with the state's research captures on the Elk Ranch, though neither agency then took management action against any bears in response to the stock losses there.[28] The Bridger-Teton Forest announced that they would review their grazing plans for their 88,700-acre portion of the study area, to potentially alleviate conflicts while not eliminating livestock from the allotment.[29]

Bear-cattle interactions increased notably in 1995. In addition to collaring more bears in their study area, biologists put ear-tag mortality transmitters on 233 calves to track their fates. Each small device issues a signal when a calf dies, allowing a researcher to search for the animal and try to identify its cause of death. Eleven calves were lost on Forest Service land, and eleven more in the park. Park rangers filed case incident records that sometimes included a special "Livestock Depredation Form" with information on the predator's bite and claw-mark locations, spacing, and other skeletal damage to the cows.

The report asked for the percentage of the carcass consumed, whether it was moved or cached, and whether there was evidence of a struggle. On the back of the form were silhouettes of a cow upon which the ranger could indicate wounds to the victim, such as "bite marks to the withers and left side of face, with one canine puncture wound into the left eye orbital and into the brain. Claw marks were visible on the back and hips, right side."[30] Several of the kills were tagged calves reported to weigh 275, 300, and even 350–400 pounds—a substantial payoff for a bear, and a significant contribution to the 500+-pound capture weight of primary adult male culprits.

FIG. 10. Cattle depredation form from Case Incident Report 972400, grizzly bear predation on calf at Elk Ranch, August 12, 1997, GTNP (Lands Records—Grazing), series 003.01, box 1, file unit 5.

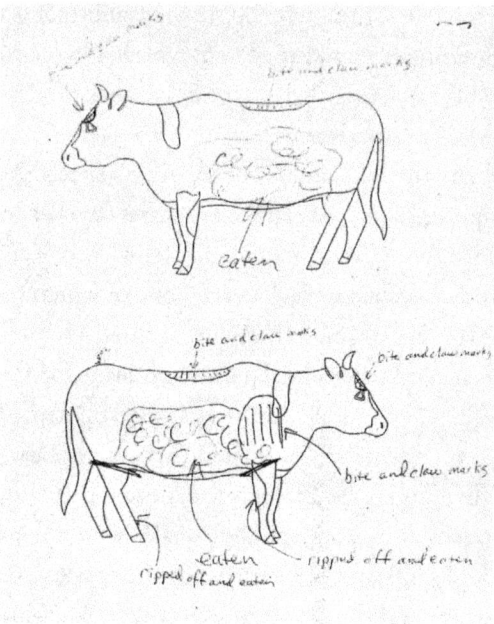

Because of heightened conflicts that season, the U.S. Fish and Wildlife Service and Wyoming Game and Fish Department again proposed relocating cattle-killing bears from the Blackrock–Spread Creek area—even in Situation 1 habitat. Eighty percent of the cattle were killed by just three grizzlies in 1995, including bear #209, who after being caught on the Elk Ranch in early September was once more translocated to the Lamar Valley of northern Yellowstone.[31] Several environmental groups objected to the move, with the Sierra Club Legal Defense Fund saying, "It's highly inappropriate even in national parks that the needs of grazing cows come before those of one of our few remnant grizzly bear populations." The *Jackson Hole News* reported that park staff were at the capture scene but that "top Grand Teton National Park officials did not know of 209's removal until after it happened."[32] That statement, whether correct or not, might have been a red flag to park managers. Nearby, Forest Service permittee Walton and the state disagreed on the number of confirmed cattle kills and associated compensation he might receive. As cattle left the summer allotments, local opinions from Jackson Hole and beyond varied as to whether managers should exercise greater flexibility in managing bears and livestock, and whether lawsuits would follow to protect the grizzlies.

By the last summer of the state's special research project, biologists had captured a total of eighteen grizzlies (twelve males and six females) and identified the presence of two other females with cubs. They also caught and monitored seventeen black bears, none of which were implicated as cattle killers. In three years researchers documented a total of 262 dead cattle, found 193 of them, and determined a cause of death for 182. Combining confirmed necropsies and assumed calf losses based on depredation rates applied to missing calves, researchers estimated that grizzly bears had killed eighty-four cows, calves, and bulls. Biologists identified the culprit in forty of forty-seven fatal attacks and, although adult males were the most frequent predators, bears of all sexes and age classes killed livestock. Grizzlies #203 and #209 killed sixteen and fifteen cows respectively, #34 killed five, while two other adult males, one adult female, and a subadult female killed one calf each.[33] (Bear #34 had only lived through year one of the study, having been killed after an encounter with a hunter in late 1994.)

In 1996 Grand Teton National Park broke a record for the number of grizzly bears captured for management reasons, though only one was for livestock depredation. After bear #203 killed a calf at the Elk Ranch, the park took no action toward him. But bear #209 had returned from Yellowstone—not a large distance for a male grizzly to cover over the better part of a year—and killed eleven calves in the last two weeks of July. News reports that traps had been set in the park to potentially destroy bear #209 did not sit well with all valley residents: the Jackson Hole Alliance for Responsible Planning acknowledged that the removal was legally allowed but, like the Sierra Club, preferred that the national park provide refuge for the bears.[34]

That was not the only controversy. Former assistant superintendent Melody Webb, in her autobiography, *A Woman in the Great Outdoors*, tells how she felt blindsided receiving a call from a reporter asking about the park's trapping and killing a grizzly on park land. (The book reports it happening in September 1995, though it appears she confused her years; park records say the bear was trapped on August 4, 1996, and subsequently euthanized by Wyoming Game and Fish officials in Lander, Wyoming. Mark Bruscino was the state biologist who caught #209 for the last time.) According to Webb, neither she nor then–chief ranger Colin Campbell knew of the incident until they followed up with park staff who were aware that #209 had killed multiple calves over the summer but had not passed that information up the supervisory chain. She wrote that the state Game and Fish agency decided to kill the bear without consulting park managers. Webb also mentioned the public outcry and said one state staffer thought that in the ecosystem, "park" meant only Yellowstone—implying that consultation was not needed with Grand Teton National Park, where the state had traditionally exercised a prominent role in managing fish and wildlife. After talking with her at the fall bear managers' meeting, Wyoming Game and Fish Department representatives reportedly apologized to Webb because she had been left out of the decision-making and assured her that the state would do better in consulting the park over future bear management actions.[35]

Steve Cain, who by then had been Grand Teton National Park's lead wildlife biologist for seven years, was out of town when the bear was trapped, though he remembered a major fuss over the whole episode. He said that state biologists had coordinated a great deal with him and park rangers. In fact, District Ranger Mel Denton had been frustrated that, although his employees regularly documented cattle losses, Game and Fish personnel needed to confirm the kills because, under Wyoming law, the state compensated owners for livestock damages and would be called to defend any appeals by ranchers. Cain also said Superintendent Jack Neckels had given approval for the state to trap and ultimately "take" (kill) the major livestock-killing bear to balance legally permitted in-park grazing with overall conservation goals. Bruscino also asserted that the U.S. Fish and Wildlife Service had authorized all grizzly bear relocations and removals.[36] Park records document that Cain consulted state biologist Dave Moody and Grizzly Bear Recovery Coordinator Chris Servheen after bear #209 renewed killing calves on July 15, 1996; all concurred that if a facility could not be found in which to place the bear once captured, it should be euthanized.[37] Another follow-up report tells the chronology of trapping efforts and mentions that, not unexpectedly for a large male bear, attempts to find a zoo or research facility were unsuccessful.[38] According to the park's press release of August 4, 1996, the three agencies jointly made the decision to remove the nine-year-old, 550-pound boar according to Interagency Grizzly Bear Guidelines. Superintendent Neckles was quoted as saying, "It is sad when any bear, especially a grizzly, has to be removed from the population. However, the loss of this particular bear will not affect the overall grizzly bear recovery efforts."[39]

Cain offered added context for why cooperation with local grazing constituents was so important to the superintendent. By 1996 the park was dealing with the recent unexpected passing of Mary Mead and, two years earlier, the death of Jeannine Gill. Both events should have ended longtime grazing privileges for the Mead-Hansen and Porter-Gill families in keeping with provisions in the 1950 park legislation. However, a major concern of park managers and community leaders at that time was the loss of undeveloped acreage as ranchers increasingly found it more profitable to sell their land for subdivision into residential lots. So Neckels, supported by the Wyoming congressional delegation, extended the two families' grazing permits to give the park time to complete an Open Space Study and recommendations to the U.S. Congress on the future of livestock grazing. The increase in private land development, and associated habitat loss, was a growing issue across greater Yellowstone. Meanwhile Jackson Hole struggled with the decline in traditional ranching. Though it was not a highly profitable way of life, some people thought of ranches as an integral part of the valley landscape.

Still, one can imagine some head-scratching on the part of the public and other NPS employees like me, then working in Yellowstone, to hear that inside Grand Teton National Park a grizzly bear had been killed for preying on cattle when neither species was commonly associated with the park. Managers from the involved agencies took their share of criticism. Conservation groups and prominent locals, such as John Craighead and medical doctor–turned grizzly bear advocate Steve French, stepped up efforts to educate residents and visitors about bears in Jackson Hole and how to live with them.

The Bridger-Teton National Forest proposed that their permittees would promptly remove cow carcass attractants and try to herd or frighten bears away from allotments, such as by firing warning gunshots, although relocation or removal of predatory grizzlies from Situation 2 habitat would still be permitted.[40] Wildlife managers hoped that targeted removal of a couple of male bears would stem the livestock losses. The national press and threats of legal action aimed at the lesser-known park in the ecosystem—seldom in the news for bear-human conflicts of any type—surely resulted in the NPS and the Wyoming Game and Fish Department being more sensitive to future removal of a grizzly bear from Grand Teton National Park.

Early in the summer of 1997, the remaining dominant cow-killer, eighteen-year-old bear #203, struck again on the Walton allotment, prompting efforts to trap or harass him.[41] Though the park itself also experienced three calf kills by an unknown bear over five days in August, the staff took no action. Bear #203's radio collar ceased functioning in the latter half of the year, and he was not heard from again; his fate is unknown.[42] Between his disappearance and the removal of bear #209, livestock losses declined to a relative trickle. By 1998 the Forest Service had approved their new plan to address livestock grazing and reduce bear predation. Wyoming Game and Fish Commissioners adopted a new formula to reimburse ranchers for animals lost to predators, a compromise between using agency records of documented kills and information reported by stock growers.[43] Grand Teton Park experienced no cattle losses that year and only three grizzly bear attacks on cattle in 1999, two of which were successful.[44] Workers installed an electric fence around the Elk Ranch pasture to deter additional attacks.[45] In 2000 a bear killed two more calves there, the last of forty-three known grizzly-cattle depredations in the park.[46]

In summarizing their work in the Blackrock–Spread Creek area, biologists wrote of similarities to and differences from what Adolph Murie had found a half century earlier. In both studies, adult male grizzlies did the most cattle killing, usually selecting juvenile cows as victims. In neither study were any stock confirmed killed by black bears, despite their presence in grazed areas. While Murie had found daytime and nighttime kills, Anderson's team found that grizzlies killed only at night, possibly influenced by greater human use of the area in the 1990s. And while Murie suspected (without radio-collared bear data for confirmation) most resident bears of being cattle killers, the later study found a smaller percent of depredators and expected that managers' hazing and translocating those bears would provide cattle owners only temporary relief. Researchers in the 1990s study did support selective removal of individual animals known to kill stock, concluding that cattle owners should not have to endure excessive losses or abandon all areas occupied by grizzly bears, as that would only heighten animosity toward the animals and erode support for population recovery.[47]

At the end of the twentieth century, the state of Wyoming debated whether to cease funding grizzly bear management outside national parks, with some supporters hoping that would prompt the federal government to remove the species more quickly from the threatened list. But the state ultimately stayed in the action, while environmental groups continued to press the U.S. Fish and Wildlife Service to improve habitat protection under the species' recovery plan. Attention on livestock depredation turned to the Gros Ventre drainage and farther south to the upper Green River Basin area, where grizzlies continued to expand their presence. In a feature article in the *Jackson Hole Guide*, ranchers Brad and Kate Mead discussed the uncertain future of livestock grazing in the wake of their changing family circumstances, the Open Space Study, and pressures from conservation groups who wanted to see less, if any, cattle in the park.[48] Cowboy Terry Schramm, a regular attendee at ecosystem grizzly bear managers' meetings in the 1990s, spoke pessimistically about the future of cattle grazing in the face of predator pressure and the challenges of managing endangered species that kill livestock.[49] Paul Walton had passed away in mid-1998 and, for many reasons, livestock grazing in and near Grand Teton Park continued to decline.

In 2004, my first summer working in the park, staff kept a few stock for backcountry use. Though rangers seldom patrolled on horses, the trail crew, and later even weed-control crews, pastured working stock on Mormon Row and near the Taggart Lake trailhead on the Inside Road. The Mead family had voluntarily chosen not to graze cows in the park for a couple of years while they considered selling part of their land. The Gill family's Jackson Hole Hereford Ranch grazed several hundred cattle that summer, driving the cows up Spring Gulch Road to a long-standing holding pen at the southeast corner of the junction between the Gros Ventre Road and highway 287. From June into July, their cows grazed reduced acreage near Mormon Row and Blacktail Butte. Firefighters had removed some of the previous pasture fencing to battle an August 2003 wildfire south of the butte, but the park postponed and eventually canceled a post-fire project to replace the fence because of the uncertain future of grazing there. In mid-season, wranglers drove the cows northward to the Elk Ranch, where they experienced no depredations from any species. I heard nothing from either staff or ranch owners about the earlier or current threat of grizzlies. As outlined in the biological assessment submitted to the U.S. Fish and Wildlife Service in 2005, prior to the reissuance of grazing permits, the park required one herder for every five hundred head of cattle as well as monitoring and prompt reporting, investigation, and removal of depredations on livestock allotments.[50]

As Jackson Hole land managers were embroiled in a multiyear effort to complete a plan for bison and elk on the National Elk Refuge and the park, the state was particularly concerned about potential brucellosis transmission from bison to cattle. Ironically, in July 2004, a group of cattle pastured south of town tested positive for exposure to the bovine disease. Thus, Wyoming required "depopulation" (lethal removal) of entire herds owned by the Gills and a neighboring rancher.[51] For the Hereford Ranch that meant cessation of their long line of specially bred cattle. The Meads sold a substantial portion of their family ranch "base" lands on West Gros Ventre Butte in 2005, terminating their long-held in-park privileges. It took years for the Porter-Gill estate to rebuild their herd, and other factors, apparently related to potential sale or alternate use of their base land, resulted in the family declining to use in-park grazing from 2005 through 2008. After Superintendent Mary Gibson Scott and I consulted with the regional solicitor's office in Denver, the park sent a letter to the ranchers in early 2009 terminating their in-park grazing for non-use, in keeping with the legislative provisions. The Open Space Study died a quiet administrative death, and no superintendent ever sent a recommendation to Congress on grazing, as circumstances essentially rendered moot the question of extending privileges for the last of the old ranching families outside Grand Teton National Park.

Meanwhile wolves, established in Jackson Hole since 1998, caused several cattle losses on the Pacific Creek allotment in 2004 and 2005. The rolling mix of meadows and forests with creeks meandering through it is high-quality bear habitat with continuing potential for depredations by wolves or grizzly bears. In comparison, the irrigated, almost treeless fenced Elk Ranch pastures are surrounded by roads, making it safer and easier for permittees to monitor the area and guard cows when predators are present. So in 2009, since the Elk Ranch allotment would be vacant of external ranch grazers, the park worked with Bridger-Teton National Forest supervisor Carole "Kniffy" Hamilton to offer Ernie Cockrell, owner of the Pinto Ranch, the opportunity to move his cattle from the unfenced Pacific Creek allotment straddling the park-forest boundary.[52] A national forest allotment can be closed only at the request of the permittee, and Cockrell agreed that in exchange for moving his cattle to the park pasture, he would ask the Forest Service to retire his permit so the free-range allotment could eventually be permanently closed.[53] The park committed to permitting him the equivalent number of "animal unit months" (AUMs: 976) that he had been granted on the entire Pacific Creek allotment, since biologists believed that the small number of cattle (about 250) in a more manageable area would result in minimal, if any, predator conflicts. By 2011 the Pinto Ranch had moved their grazing operation to the Elk Ranch, and later they converted from having cows and calves—a more tempting predator target—to summering only yearling steers.

By 2019 just 23 horses owned by the park and 184 by the Triangle X and Moosehead Ranches grazed park land. As for cattle, the last surviving Wilson brother still, as the family had for decades, paid a token fee through the National Park Foundation for his lifetime lease from John D. Rockefeller Jr. to graze three dozen longhorns near Kelly; there is no provision for that lease to pass to another generation. Nearby, in the Gros Ventre drainage, there was renewed discussion of the state of Wyoming needing to maximize the value of the last 640-acre state "school section" in the park east of Kelly. (In the 1785 land ordinance that resulted in western lands being surveyed into townships and sections, two sections were reserved for the maintenance of public schools.) The parcel was estimated to be worth more than sixty-two million dollars as of 2023, if subdivided into thirty-five-acre lots overlooking Jackson Hole—and it reaped less than $3000 annually, funds used to benefit schools, when leased by the state of Wyoming for cattle grazing. Strong public opposition to auctioning the land prompted the legislature, in February 2024, to authorize a direct sale to the federal government for not less than one hundred million dollars. Apparently without much objection, the law as passed would require the park to continue public hunting and livestock grazing, as the state has traditionally permitted, on this acreage.[54]

For those concerned about livestock and associated predator conflicts within Grand Teton National Park, there has been considerable progress since the park's expansion. Over six decades of evolving circumstances, permitted livestock grazing was reduced by 89 percent, and all grazing privileges associated with ranches outside the park ceased.[55] The Pinto Ranch, the last private inholder with cows, chose not to graze park land in 2020–2023, although the owner did so again in 2024. The Pacific Creek Forest Service allotment remained vacant, and its formal retirement awaited completion of required environmental compliance.[56] And, as part of long-term efforts to remove or modify fencing to be more wildlife friendly, park staff and volunteers removed the former cattle holding pen at Gros Ventre Junction along with other unused fencing in the Antelope Flats and Elk Ranch areas. No bear predation on livestock had, as of 2024, occurred in Grand Teton National Park since the turn of the twenty-first century despite there being far more grizzly bears. As minimal livestock graze each summer, bears killing cattle in the park is, we can hope, a thing of the past, as managers have increasingly turned their attention to other challenges involving bears and humans.

Wake-Up Call for a New Century

On a warm, stunning day in August of 2004, my first summer in the Tetons, my husband and I, our daughters off in the care of friends, hiked up a forested trail and beyond to Garnet Meadows, best known as a backcountry campground used by climbers bound for the summit of "the Grand." I confess to never being bitten by the climbing bug, partly due to an old back injury (obtained by falling rather ingloriously from a ladder); my knees have long been mishap-weakened as well. The "climb" through a large boulder field some four miles up the route is sufficient challenge for me, and afterward we were pleased to sit overlooking the campsites dispersed among large granitic rocks in the alpine meadow. From there, one can look up toward the narrow trail near Spaulding Falls where there are few trees among the subalpine vegetation. We marveled at the Middle Teton, centered above the meadow, and watched hikers disappear out of sight along the most common route to the park's tallest peak. I could hear the creek cascading below us into the canyon. More often, I was struck by the sound of falling rocks competing with the occasional chirping bird.

As we sat eating our lunch, a park ranger, in camp while awaiting his time for summit patrol or rescue response, briefly joined us. I looked out over sites occupied by campers' tents. In several places long sticks, each with a small bundle or pack tied to the top, leaned against boulders and rock walls, and I asked the ranger what that was all about. "That's for food storage," he said. "Hikers leave their food while they are off practicing their rock climbing. That way the mice and the marmots can't get it." I asked about bears, and he said they never had bear problems at the meadows. Though I was too shocked to say so out loud, I'm sure I thought, "never say never."

The new century saw an average of 2.5 million or more visitors to the valley of the Tetons each year. It also brought more grizzlies to the park and, only coincidentally, me, who more than once was teasingly accused of leading them southward from Yellowstone. Bear management was not yet the major issue that it had been for decades in my prior park, and my early impression was that it was not as common for Grand Teton employees or visitors to store backcountry camp food in bear-resistant canisters or high on tree branches, to carry bear spray, or to be too concerned about potential bear encounters. In a way typical of the frequent time lag between scientific discoveries, their (mostly technical) publication, and common knowledge, National Park Service and other publications continued to say that grizzly bears were "occasionally seen in the remote northern corners of the park." It seemed to me that many locals and even some park staff and managers were stuck in the 1980s, just unaware of or in denial about the species' recovery in their backyard, despite the fact that some of them had worked on the recovery effort, at least to a degree, for two decades. Park records, and certainly the history of cattle-killing grizzlies, evidenced the bears' increase. By 2002 the Grizzly Bear Study Team had mapped grizzly bear females and cubs (which typically follow the geographic expansion of subadult and adult male bears) living in the northern third of Grand Teton National Park, and two years later their range had pushed even farther southward.[1] In retrospect, this should have put the park on higher alert. Then again, are either bureaucrats or average citizens ever good at foreseeing change on the horizon? Even I, being responsible for all the science programs in the park, consistently struggled to promptly read newly published data on any subject when the busyness of today was always calling for attention. Did I and others need to see grizzly bears in the valley to believe they were there, to be ready to deal with them?

While I have not kept meticulous track of the countless wildlife sightings in my life, I have noted many of them in my journals, which also reveal the relative *absence* of observations, especially of either bear species. This occurred in my first few years in Jackson Hole. A few black bears damaged property each year from 2001 to 2005 (see table 1, chapter 4). I also recorded an incident when several summer employees, sampling vegetation off-trail in Moose Basin, had a surprise close encounter with a grizzly bear in late July 2005. It was all over in fifteen seconds and no one, bear or human, was injured. Then, while driving through the park on Friday evening of Memorial Day weekend 2006, my family and I saw a nice-sized radio-collared grizzly bear mother with three cubs-of-the-year crossing the fast, shallow water of gravelly Pilgrim Creek. It was my first sighting of then-ten-year-old bear #399, who at that point was scarcely known to park staff or visitors. This was not long after my call from the ranger at Colter Bay, mentioned in the opening to this book. The director of the in-park day care center where I dropped off my youngest daughter each day had also chanced to see mama bear and cubs near her home in the northern housing area. She had excitedly shared her observation with the children, who relished seeing the bears for themselves. In late June, because of the grizzly family's increasing visibility near park roads and facilities such as the Jackson Lake Lodge, staff from varied park departments met to discuss the increasing workload of "bear jams"—traffic jams associated with the animals, something that had occurred for many years in national parks such as Yellowstone and the Great Smoky Mountains. Grand Teton visitors had often stopped to watch roadside black bears since the 1960s; moose, elk, and bison jams were also common. But more and more, field staff were being dispatched to manage such traffic, diminishing their ability to be on time to staff visitor center information desks, to lead walks, or to respond to visitors needing assistance, even emergency help. Despite long-standing guidance in place about standard bear management practices and safety measures, workers wanted clearer direction on how to deal with bears in and around campgrounds and picnic areas. I could hear the unspoken concern—*now, there are grizzlies.*

In mid-July, I joined Steve Cain, a highly competent man with a perpetually calm demeanor and considerable experience managing bears and other animals, at the chief ranger's staff meeting. It was reported that lax food storage in developed areas had attracted two black bears that, as we classified it, "received food rewards." This is a dangerous precedent that often leads to aggression by those bears against humans or property. Some rangers, who shouldered a fair amount of the burden of responding to the increased number and visibility of bears, encouraged us to prepare a new bear management plan. But we advised sticking to the fundamentals already in Grand Teton's strategy: educating people about bears and keeping food attractants unavailable. We emphasized the need to better communicate park regulations and "best practices" to visitors and employees, especially to concessions workers, who had recently taken over the responsibility to manage campgrounds previously run by the Park Service.

Rangers floated interesting ideas for getting "more tools in their box." One suggested that we use aversive conditioning or "bring in dogs and get the bears out," alluding to Karelian or other trained bear dogs that are used to chase bears away from human-occupied areas, notably at that time in Glacier National Park. I had watched experiments using dogs to push bears out of human-occupied areas at Yellowstone and in Canada's Banff National Park, and these animals can be effective. But it is tough to truly "condition," or teach, bears to avoid easily available food temptations, whether natural or human-provided. With intensive and repeated effort, you may succeed at deterring individual animals from specific locations, rather like the way you train your dog to "do its business" in a certain area of your yard and not in another spot. Teaching them to avoid using the entire yard, or to not eat from another dog's bowl if offered the hungry chance, is a bigger challenge. In the end, rather than invest in bear dogs, Yellowstone permitted specially certified rangers to use rubber bullets, bean bags, and noisemakers to haze bears out of campgrounds and housing areas—something that we also permitted at Grand Teton. But Steve and I both strongly felt that to benefit grizzly bear recovery and visitor opportunities we should follow our sister park's practice of managing people, allowing bears access to as much good habitat as

FIG. 11. Rangers manage bear jam, Pacific Creek Road. National Park Service photo.

possible—including areas along roadsides and adjoining developed areas. Of course, this would still require a staff- and time-intensive program.

Since we were well into the summer season, the year's budget was already programmed, and hiring more staff was not doable in a useful time frame. I promised to provide extra "on call" staff, paid for with the small amount of discretionary funds I had at my disposal. Seasonal ranger Kate Wilmot, who had previously worked bear patrols at Glacier, had expressed willingness to work overtime helping with bear jams, and graduate student Leslie Frattaroli also agreed to help when she was not busy with her field research. The two of them bolstered the staff's capacity to handle wildlife jams for the rest of that season—not a long-term strategy, but it bought time for the park to ramp up its ability to respond to more bears in more places.

At a September resource information exchange, the keynote speaker, Grizzly Bear Recovery Coordinator Dr. Chris Servheen, told staff that the grizzly bear population in the Yellowstone ecosystem had doubled every twenty years since the 1940s (though it should not be expected to continue

at that rate); they were here to stay in the Tetons. In 2006 the park experienced a relatively small number of bear-human conflicts, and some visitors
and staff were fortunate to see a grizzly mother with cubs-of-the-year, a rare
treat. It was not just #399 that prompted a new focus on bear management in
Grand Teton National Park, but certainly her visibility and burgeoning popularity, the expansion of grizzly bears into Jackson Hole, and the black bears
that had long and consistently occupied all suitable habitat struck me like
the bullseye on a target. The park had a sound, if dated, bear management
plan implemented by a dedicated and experienced staff of rangers, biologists,
and maintenance workers. And so we would aim for improvement, from the
strategic to the mundane; the staff was up for, and to, it.

As the snows of autumn began to accumulate on the peaks and park
roads, and facilities closed for the season, we launched a major review of the
bear management program. Our goals were to clarify the responsibilities of
partners such as concession companies and of workers in all disciplines; to
update the public and staff through information and outreach; to strengthen
enforcement of regulations; to ensure that research and monitoring support
restoring and maintaining natural integrity, behavior, and distribution of
bears; to provide for human and animal safety; and to examine the park's
combined capacity to address current and future challenges. In the plus column, the park had promoted educational messages about bears and required
"proper" food storage and disposal since the 1970s as did other "bear parks."
In 1989 only about 50 percent of the trash receptacles in the park and parkway
were "bear-proof" or, more accurately, "bear-resistant"—reflecting peoples'
limited abilities to outsmart bruins. But by 2006 nearly 100 percent of the
existing trash cans and dumpsters were, and most of them in good working
order. Best of all, there was a very low rate of bear-caused human injuries and
only one human death associated with a bear (see tables 2 and 3, chapter 12.)

However, staff saw problems and areas for improvement: Increasing numbers of day hikers and backcountry campers encountered (still mostly black)
bears. Some food and garbage was improperly handled on inholdings and
adjacent lands, and compliance with and enforcement of food and trash storage regulations was inconsistent in both developed front-country areas and

the backcountry. Protocols on when, where, and how quickly to report bear sightings or food storage infractions were unclear. We also faced the ever-lively question of whether and where to tolerate animals that had become used to humans, the continual need for effective training of park and partner staff, and the growing number of time-consuming wildlife jams. Also, messaging at entrance stations, visitor centers, and other facilities and on the pre-trip planning website was insufficient, inconsistent, and outdated—the park was still using decades-old informational brochures, signs, and messages. Managers thought a fresh approach to the whole program was needed and, unlike other large parks with the animals, Grand Teton National Park had no dedicated bear management office or staff, even seasonally. By the time of bear emergence the following year, the superintendent wanted to start making changes, especially by dedicating more people to deter and manage bear-human conflicts.

Officially, a "Wildlife Brigade" began in the summer of 2007, with seasonal ranger Kate Wilmot dedicated to recruiting, training, and leading three volunteers and one other employee who promoted safe behaviors for watching wildlife and living or recreating in bear country. The brigade was popular with visitors and employees, who were relieved to have extra help, and the park's management team was in favor of adding a new staff position to continue the effort. But as with most government programs, newly identified needs had to compete for and, even if viewed as priorities, were unlikely to receive more money for a year or two at least. So, behind-the-scenes work included the tedious preparation and submission of budget proposals into what could seem like a black hole from which the light of funding approval might never emerge. Luckily, parks received a Centennial Challenge call that year for additional staff positions. The chief of visitor and resource protection ("chief ranger") and his budget analyst prepared a proposal to fund a new permanent ranger who would lead the Wildlife Brigade and be dedicated to bear management. Wildlife biologists would continue to oversee plans and bear captures or removals when needed.

Meanwhile, the park rolled out a new "Be Bear Aware" educational campaign, borrowing the phrase from neighboring national forests. Workers installed large signs near park and campground entrances alerting visitors that Grand Teton Park was bear country. The staff graphic designer developed catchy versions of plasticized posters for bulletin boards, restroom walls, and other locations informing visitors of how "You Can Save a Bear" while camping or hiking in the park. Bright yellow diamond-shaped stickers with bear paws and food storage messages went on trash dumpsters and picnic tables.

Biologists revised guidance on when and how to report bears and offered training sessions for both park and concessions staff, since workers at restaurants, hotels, and other services outnumbered rangers and encountered as many visitors with wildlife questions. The park's concessions chief worked with the Grand Teton Lodge Company on multiyear funding to place bear-resistant food storage boxes in commercially operated campsites. Backcountry campers in Grand Teton, who, unlike in neighboring Yellowstone, do not find designated campsites with food-hanging poles already installed, were newly required (except in locations, such as Cascade Canyon and lakeshore sites, where there were bear boxes) to carry bear-resistant plastic food storage canisters that the park provided on loan. This requirement was initially only for those camping below 10,000 feet in elevation, since few people reported seeing bears above that; after a few years of backpackers' acceptance, the park changed the rule to apply across the high country.

Although the location and condition of septic systems was "not my department" and had not been on my mind in my first years at Grand Teton, an employee had clued me in to a collapsed fence around the sewage treatment ponds not far from the concessioner's horse corrals at Colter Bay. No one recalled a bear getting into the lagoons, but I implored maintenance staff to repair the fence around the site, mentioned years before as a potential bear attractant. I was grateful that workers responded quickly, though I did have to stifle a laugh when I toured the site late in the summer. The fence around the lagoons had indeed been replaced with a tall and sturdy mesh enclosure, anchored into the ground to deter bears or other wildlife from digging under it. The locked entry, through which employees drove to maintain the site, was spectacularly *not* bear-resistant—just a standard swinging gate that a

FIG. 12. Black bears get food along lakeshore in Grand Teton. National Park Service photo.

bear (or person) could easily climb over, under, or through. When I reported that to the chagrined maintenance chief, he said his staff would get right on it. It reminded me that when checking facilities and locations, fresh eyes (and noses) are constantly needed, through which observers try to sense like a bear, ever on the search for what could be food.

Ironically, and in the end perhaps fortuitously, Grand Teton experienced one of its busiest seasons for bear activity in 2007.

The first serious bear-human encounter in many years occurred when grizzly #399, with her then-yearling cubs, injured a visitor. Several black bears that season repeatedly demonstrated aggressive behavior, approaching campers and frightening them from their picnic tables at mealtime. After staff removed four black bears from the park's population for posing an unacceptable danger to humans, Superintendent Mary Gibson Scott, who took a personal interest in wildlife, especially bears, was frustrated by the number of animals in trouble, though Steve Cain assured her that areas outside the park were experiencing a bad year as well and that, despite these arguably sad control actions, the bear population would not suffer from those removals.

Mary fully supported expansion of a dedicated bear management program for the park and the Rockefeller Parkway. She was instrumental in reaching out, through the Grand Teton National Park Foundation, to private philanthropists, who provided financial support for new efforts. By 2008 the park had identified the goal of having a bear box in each of the estimated one thousand front-country campsites and high-use picnic spots. The beauty of such a food locker is that visitors can safely stash their food, cooking gear, and other bear attractants while not immediately in use. A ranger or campground worker on patrol who finds unattended food items can move them to the bear box and leave a courtesy notice reminding visitors that even water bottles or a six-pack of beer can draw a bear's attention. In the second full year of the "Be Bear Aware" campaign, the foundation launched a program to fund the purchase and installation of more bear boxes. This would significantly augment funds available from the government and the Lodge Company's project.

The popular Wildlife Brigade grew in 2008, with three seasonal rangers, two interns funded by the nonprofit Greater Yellowstone Coalition, and seven volunteers who came from around the nation to work three or four days each week to educate visitors and encourage safe wildlife viewing. Most importantly, Park Service funding for a new ranger had come through, the government wheels spinning more quickly than usual. As originally envisioned, the park would gain a commissioned employee capable of enforcing laws and regulations under supervision of the chief ranger, who at the time was Andy Fisher. He and I had been stationed at Yellowstone, including one winter we worked together at Old Faithful, more than two decades before. Andy had taken part in bear management actions and was very engaged with Grand Teton's program review. But evidently he was unaware that the newly approved staffer would be responsible for recruiting and overseeing a growing cadre of brigade volunteers. A worrisome late-winter search and rescue operation prompted the chief ranger to resist adding more people to his already sizable staff—some sixty permanent and two hundred seasonal employees and volunteers. By mid-April grizzly bear #399 and cubs emerged from their winter den and got into a bird feeder at a private home just east of the park boundary. Wildlife jams had begun even before summer employees had arrived, and the park needed focused personnel with experience managing human-wildlife interactions, fast. When I offered that the Science and Resource Management Division would gladly take responsibility for and advertise the new bear management position, albeit without the law enforcement commission, which only he could oversee, Andy seemed relieved.

At the time, most new permanent positions at Grand Teton and other parks were "subject to furlough" to allow funding flexibility—full-time employees had to be wholly "base-funded" using congressionally appropriated ONPS (operations of the NPS) funds. Furlough positions required only six months' base funding; added pay periods could be funded using recreational fees and entrance station revenues, concessions franchise funds for special projects related to the positions, even donated funds from the Grand Teton Foundation. As Steve Cain and I went to chat with the Ranger Division's budget analyst about transferring newly appropriated funds for the bear manager to our division, I recall both of us expecting that we'd receive just enough money for a half-year employee—the length of the bear-viewing season—and wondering how we could augment that for a few months more to do program planning, hiring, and recordkeeping. Imagine our surprise when the budget analyst told us, apologetically, that she and the chief ranger had written the proposal to cover "only" nine or ten months of salary. Yet funds were enough for a journeyman-level commissioned park ranger (whose pay rate is higher than for non–law enforcement staff). She would transfer nearly twice as much money as we had anticipated. It was one of those close mouth, nod, and say, "That will be fine, thanks!" moments in my career. Steve and I returned to our offices laughing that at least we shouldn't be accused of padding our resource budget requests; more likely we were underestimating the true cost and fundability of our proposals. Later that summer we officially welcomed ranger Kate Wilmot as Grand Teton National Park's first permanent employee dedicated to bear management.

Two other items of note progressed in 2008. The flurry of funding requests made the previous year resulted in another success: $71,000 from the NPS Natural Resources Preservation Program for a two-year project to further develop the Wildlife Brigade of volunteers, evaluate the success of their efforts, and study the effectiveness of the "Be Bear Aware" information campaign. The latter part of the project would be accomplished through a cooperative agreement with the University of Wyoming's Survey and Analysis Center to survey visitors in 2009. And, although I had stayed out of the discussions about backcountry food storage requirements, biologists were frustrated with their inability to convince field rangers that the climbing community and their packs were not immune to attracting bears. At dispersed locations in the Tetons, it was common to see, as I had, day or overnight users' packs leaning against rocks or hung on easily reachable tree branches at the base of attractive climbing routes. Rangers argued against installing infrastructure, including bear boxes, in the de facto wilderness zone. While I consider myself a true supporter of wilderness values, I struggled to balance those with what I and my staff considered the inevitable attraction of bears to unsecured food in the backcountry. Finally, in a discussion with the superintendent, biologist Cain confessed that he had tried to no avail to convince rangers of the need for placement—without permanent installation in concrete, as is done in roadside campgrounds—of food storage boxes in Garnet Meadows. So at Scott's direction, the park helicopter crew flew boxes up Garnet Canyon late that season. When I retired eleven years later, although wilderness rangers periodically advocated for their removal from the backcountry campground, the food lockers remained. As of 2023 no property damages or human injuries had been documented in the climbing zone, although bear-human conflicts had occurred throughout the backcountry, including both forks of Cascade Canyon and lower portions of the popular trail from Lupine Meadows.

In the first decades of the twenty-first century, Grand Teton National Park focused on more wholly contributing to the evolution of bear expansion and recovery into the southern portion of the Greater Yellowstone Ecosystem, while aiming to make it safe for staff and a growing number of visitors to coexist with bears in Jackson Hole. The park and parkway finally were recognized as places to see both grizzly and black bears, areas with a substantial visible effort to manage the "bear-human interface" using uniformed rangers and volunteers. The record of wildlife-people conflicts is low, for a park with three or four million visitors each year, but not without some major incidents that speak to the power of bears.

12

Bear Danger

In the spring of 2018, as I was nearing the end of my time with the National Park Service, my boss asked me to escort a visiting dignitary on an evening wild-life watch. The guest was highly placed in the Department of the Interior, on an official visit to discuss Moose-Wilson Road improvements and the budget rather than wildlife. But as often happens, the superintendent wanted the man, stationed in Washington DC, to enjoy and learn about as many aspects of the park as possible. Dressed in my green and gray Park Service uniform and flat "Smokey Bear" hat, I picked him up at the Jackson Lake Lodge. I had high hopes of satisfying his wish to see a bear and a moose, as it was prime wildlife view-ing season and both species had been frequently seen nearby in recent days. I'd learned, however, to have a back-up strategy in case luck failed us, as it did that evening. The roadside wildlife jams that had entertained visitors all day were gone, as were the charismatic animals that had prompted them. After an hour of fruitless effort driving around, I pulled out my Plan B, suggesting we take a short walk to see nesting trumpeter swans at Christian Pond, a half mile from the lodge. He was game, so we headed out on the trail at about 7:30 p.m.

The trail meanders through sagebrush, willows, and small stands of aspens and conifers along Christian Creek then climbs a small hill to overlook the pond, where the swan pair, at least, did not disappoint. It was a warm, quiet mid-May night, and even so close to the park road we had the trail to ourselves. As we turned to head back to the lodge, which we could have seen in the distance were it not deliberately placed to sit in low profile surrounded by trees, there was movement down a small knob about seventy-five yards in front of us. "There's a bear," said my companion. "It's a grizzly," I affirmed, raising my voice louder than my normal conversational pitch: "Let's just back up atop this hill for a bit and see where it's going to go." The radio-collared bear, a subadult or young adult, moved briefly out of sight as I quietly released my bear spray from the strap holding it in its holster and slipped off the safety—something I've done only a few times in my life, on or off the job. The bear reappeared at a low spot on the trail between us and the lodge, moving toward us but not, I thought, with intent. Nonetheless, my career momentarily flashed before my eyes as I pictured the headline, "Ranger Leads DOI Official into Bear Encounter." I loudly said something, who knows what, to make sure the bear knew of our presence, and it turned off the trail and disappeared again in the undulating terrain. We stood still as I talked about nothing significant that I recall, just deliberately keeping up chatter. After a seemingly endless ten minutes or so, during which we had not seen the bear again, my guest asked, "How do we get back to the lodge?" I told him we would return the same way we'd come, and I led him, fortunately without incident, back along the trail in the fading twilight.

The next morning, before I could report the close encounter to my bosses, I received a text from Superintendent David Vela, who had picked up our guest to drive him to the Jackson Hole airport for an early departure. Our visitor shared his excitement at having seen a grizzly bear the night before; he asked David to thank me and tell me they had also scored a moose sighting en route from the lodge that morning. David spent the next several years in the nation's capital serving as acting director of the NPS and occasionally ran into this official in the halls of the main interior building, where he reportedly often told numerous others his proud tale of seeing a Grand Teton grizzly bear.

Would that every bear sighting, along a roadside or in the backcountry, be as benign as our sighting that night turned out to be. Bears truly can pose a danger to their own kind, to other wildlife species, and to people. Across North America, the rate of conflicts—property damages and injuries—between humans and both black and grizzly bears has increased, especially during the last five decades, even though compared to other sources, bears cause a tiny fraction of human injury and death.[1] (Stephen Herrero's *Bear Attacks: Their Causes and Avoidance* is a fine comprehensive review of bear encounters in the United States and Canada.)[2] While a high percentage of the hundreds of people wounded by black bears across the continent have sustained only injuries, black bears reportedly caused sixty-three deaths between 1900 and 2009, mostly in Canada. The less numerous grizzly bears have also injured hundreds of people where they range in North America and killed nearly eighty individuals since 1900.[3] In the ecosystem, bears have killed eight people inside Yellowstone National Park and another twelve outside park borders, most recently in July 2023.[4] These incidents, and the ever-present danger of bear-human encounters, understandably prompt wildlife managers to spend considerable effort to keep people and bears at safe distances.

None of the fatalities listed for the ecosystem is attributed to a bear in Grand Teton National Park, though during my tenure there we lacked a comprehensive list of local bear attacks on humans, and this new compilation suggests one bear-related incident should be included. Park records and newspaper accounts reveal at least twenty-four encounters resulting in twenty-seven bear-related injuries over the past half century inside Grand Teton National Park and what's now the John D. Rockefeller, Jr. Memorial Parkway. Eight incidents involving grizzlies wounded eight people, while sixteen black bear incidents and one unknown species of bear reportedly injured at least nineteen people. Most incidents did not cause significant injury, and no human fatalities are directly attributed to bear *attacks* in the park and parkway as of mid-2024. However, one person died and one was injured while hiking and trying to avoid bears; another was injured while trying to chase a bear away. I have included them in the above total because the park noted them in accounting for bear incidents and actions or, in the case of the fatality, rangers' investigations linked bears to the victim. In addition, at least two grizzly bear–caused human injuries occurred just outside the park's east boundary. There have been other, more distant, bear-caused human injuries in the Bridger-Teton National Forest, particularly in the Thorofare region northeast of the park, which are not accounted for here.

TABLE 2. Reported injuries caused by black bears in Grand Teton National Park and the John D. Rockefeller, Jr. Memorial Parkway, 1929–2023.

Date	Number of Injuries	Location	Note
Summer 1965	1*	Cascade Canyon trail crew camp	Employee fell and injured leg while trying to drive bear from camp cook tent
Summer 1967	1	Along rd. ½ mile s. of Flagg Ranch	Boy injured in car with window open; suspect bear eventually shot
August 1976	1	White Grass Ranch	Bear bit wrangler trying to rope and herd it
10/9/1976	1	Phelps Lake	Bear bit female backcountry camper
June 1977	1	Cascade Canyon	Bear nipped Boy Scout in calf
Summer 1977	2	String Lake	Bear scratched and mouthed two campers
7/15/1978	1	Trapper Lake	Backcountry camper in tent mouthed, bruised
Summer 1979	2	Unknown	Listed on park's annual report, no details
8/14/1985	1*	Stewart Draw	Woman died after falling from tree she climbed to avoid bear
June 1994	2	Hidden Falls/ Inspiration Point	Bear licked boy, nipped two others on trail, severity of injuries unclear; bear relocated
7/20/1997	1	Deadman's Bar cook site	Bear bit cookout worker, was removed
7/24/1997	1*	Stewart Draw	Hiker injured avoiding bear, no detail; species in question
6/5/1998	1	Jackson Lake Dam	Bear bit woman walking on path
7/6/1998	1	Jenny Lake	Bear bit woman walking on lake-shore
7/18/2006	1	Colter Bay Campground	Bear mouthed child in sleeping bag
7/19/2006	1	Granite Canyon	Female bear w/ cub(s) tore tent, pawed person's head

*Injury/fatality is bear-related, although not directly caused by the bear.

TABLE 3. Reported injuries caused by grizzly bears in Grand Teton National Park and the John D. Rockefeller, Jr. Memorial Parkway, 1929–August 1, 2024.

Date	No. of Injuries	Location	Note
8/14/1994	1	Emma Matilda Lake	Bear severely injured jogger on trail
9/1/1997	1	Glade Creek, Rockefeller Parkway	Female with yearling cubs attacked moose hunter who had a bow near fresh elk calf kill
3/7/2001	1	Upper Berry Creek	Bear followed and wounded skier
4/15/2001	1	Snake River bridge, Rockefeller Parkway	Bear followed, scratched fisherman
10/15/2001	1	School House Hill near Moran Junction	Female with cub-of-the-year wounded hunter in elk reduction program
6/13/2007	1	Wagon Rd. behind Jackson Lake Lodge	Female grizzly with three yearlings on elk calf kill surprised, wounded man walking
10/30/2011	1	Snake River bottom	Bear on carcass surprised, wounded hunter in elk reduction program
5/19/2024	1	Signal Mountain	Bear(s) surprised hiker at close range, biting him and his bear spray can

Close encounters between bears and humans have been reported in inconsistent ways over Jackson Hole's history of more than a century. In the earliest days of settlement, historical accounts hint at a few injurious encounters in vaguely described locations. For example, in reporting upon the death of Marie Wolff, a longtime valley resident who lived north of Spread Creek in what's now Grand Teton Park, the *Jackson Hole Guide* included the story of a young man nearly killed by a bear, presumably in that vicinity. The man's "throat was torn until his tongue was exposed, his left ribs laid bare and his left arm nearly severed at the shoulder. He was taken to the Wolffs' where he was nursed all winter eventually regaining his health the following spring."[5]

Marie and her husband, Emile, had received a patent for the ranch property in 1906, and the incident reportedly happened not long after the family located there, but because the news provided no clear date, bear species, or location of the attack, it is not included in the incidents noted above.

Grand Teton archives and old newspapers contain no records of bear-caused human injuries from the park's inception in 1929 into the 1960s. As visitors, staff, and their encounters with wildlife increased, reports reveal bear-related injuries, although detail is sometimes scant. The first documented incident, a close call that could have resulted in the bear making contact with the man, though the record does not say, occurred in 1965 when a member of the Cascade Canyon trail crew fell and injured his leg while trying to chase a bear out of the crew's camp cook tent.[6] The first evident bear-caused human injury, in what is now the Rockefeller Parkway, occurred in 1967 along the main road about half a mile south of Flagg Ranch, which at that time was on the floodplain of the Snake River just north of the highway bridge. That location would place the incident near the base of Huckleberry Hill, where prior to the 1988 wildfires the area was covered with thick stands of aged lodgepole pines. Today, in the post-burn younger stands, grizzly bear sightings are not uncommon, but since the record does not specify which species, it can be assumed that a black bear caused the injury. Though nothing appeared in local newspapers, the park wrote of an unprovoked attack on a twelve-year-old boy in a car with an open window. Park rangers speculated that the bear was a known roadside beggar near Lizard Creek Campground, but since they could not trap it, they shot the suspected culprit.[7]

Despite persistent bear activity along park roads, at still-existing dumps, and in backcountry trail crew and other camps, it was 1976 before two more bear-caused human injuries occurred.[8] The headline in the local paper read "Bear Baiter Bitten" when White Grass Ranch wrangler Rusty Long swung a rope and tried to herd an animal into a trap and took a bite to his hand. "It was a crazy thing to do. It was surprising that no one was killed. That was a big bear," said ranch owner Frank Galey, who estimated that the black bear weighed about 450 pounds. Jurisdiction over management of wildlife on private land inholdings in the park, at least to non-attorneys, has often been hazy, and it is unclear whether Wyoming Game and Fish Department or federal officials placed a trap near the open garbage dump that the park had asked Galey to close or whether capture occurred. A park spokesperson said, "It's his land and he can do what he wants."[9] (A life or term estate already sold to the government, as the ranch had been in 1956, technically already belonged to the United States, though owner/seller Galey retained the rights to use the property for the remainder of his life. It was not NPS practice to treat such acquired lands as park property until the sellers had completely vacated, which at White Grass occurred in 1986.) A second incident in America's bicentennial year occurred at Phelps Lake, where a female camper received superficial wounds when a black bear bit through her sleeping bag.[10] Rangers closed Phelps Lake backcountry camping for the season after this injury.

Three reportedly minor injuries happened in 1977, when a yearling black bear in the backcountry sought food in a boy's pack and nipped the child in the calf. Later, two illegal campers were mouthed and scratched by an adult bear near String Lake.

The following year, a backcountry camper who had left food in his tent at Trapper Lake suffered minor scratches from being mouthed and bruised by a black bear.[11] The park provided no details on two human injuries reported in 1979, though rangers closed backcountry campsites at Trapper and Bearpaw Lakes due to bears, as they had the previous year.[12]

The 1980s had no bear-caused injuries but was marked by the unusual instance of the park's only (through mid-2024) bear-related human fatality. On August 14, 1985, Wilson, Wyoming, resident Susan Walker was hiking alone through the mostly forested Stewart Draw, returning from a hike up Buck Mountain in the southwestern part of the park. She evidently climbed a subalpine fir tree to avoid a black bear, fell seventy-two feet to the ground, and received such serious injuries that she died. During their investigation, rangers saw a female black bear with cubs in the area and noted bear claw marks more than fifty feet up the tree, suggesting that one or more bears pursued the woman into the branches, although the actual circumstances of the unwitnessed incident are unknown.[13]

Another nine years passed with no humans injured by bears. Then, in late spring of 1994, an aggressive black bear harassed hikers along the popular trail to Hidden Falls and Inspiration Point, obtained food, licked a child's face, and nipped two people's legs. Jenny Lake and Cascade Canyon trails, which meander through mature mixed conifers in granitic rocky terrain, were closed, and when biologists setting a culvert trap were approached by a bear, they shot it with tranquilizer darts and moved it outside the park; no subsequent black bear incidents occurred.[14]

However, Grand Teton National Park's most serious bear-caused human injury occurred later that year, and it was caused by a grizzly. Two Ocean and Emma Matilda Lakes lie not far from the eastern boundary of the park, north of Moran Junction. The lakes are similar in size, surrounded by mixed conifer and aspen forests with some open sagebrush-grassland meadows. Hikers can circumnavigate one, or even both, in a half- to all-day hike, depending on their energy level, as both trails cover rolling terrain but nothing particularly steep. A four-and-a-half-mile drive up a dirt road off the paved Pacific Creek Road brings visitors to a trailhead parking lot at the outlet of Two Ocean Lake. The morning of August 14, 1994, Park City, Utah, resident Michael Dunn was jogging on the north shore trail around Emma Matilda, between the two lakes, when he surprised a grizzly bear, which knocked him down. The bear bit and clawed the man, who initially resisted then played dead, and the bear ran off as abruptly as it had appeared. Dunn stanched his bleeding and found help from people nearby on the trail. The jogger suffered severe lacerations and puncture wounds to his head, face, neck, shoulder, groin, and leg and, after a helicopter ride out of the wilderness, he underwent surgery to have his leg muscles reattached. In later interviews, Dunn, who recovered fully, said the bear was eating berries when surprised. The park closed the area around the lakes for twenty days but took no further action against the animal.[15]

Only five weeks later, another grizzly bear attack occurred nearby, just outside the park. (Incidents outside the boundaries are not counted in the number of park injuries.) On September 20, 1994, an elk hunter had separated from his companions on Davis Hill in the Bridger-Teton National Forest northeast of Moran Junction when he surprised the animal. Clayton Peterson yelled to scare the bear and shot it at about eight feet distance. Despite being hit, the bear turned to charge and knocked the man down, biting him in the head and arm. Peterson was airlifted to the hospital in Jackson and treated before being transferred home to Casper, Wyoming. State Game and Fish officers searched for and found twenty-two-year-old radio-collared male bear #34 to be seriously wounded, so they dispatched it.[16] Grizzly #34 was one of the bears documented, during the state's 1994–1996 study of bear depredation, as having killed multiple cattle in the Elk Ranch and Spread Creek areas.

In 1997 two people were injured by bears, one a grizzly and one a black bear, with a third bear-related non-attack. A hiker in Stewart Draw—the same area in which Susan Walker had fatally fallen from a tree while avoiding a bear a dozen years before—slipped off the trail and injured his ankle while trying to avoid what he thought was a grizzly. Rangers completed a Case Incident Record for a "Bear Incident/Injury," but the species is unclear. (Because of the limited distribution of grizzlies in the park at that time, I show this incident on the black bear table.)[17] The first actual bear encounter that year occurred at Deadman's Bar, a popular Snake River launch and takeout spot where the Jackson Lake Lodge concessioner has a site to provide meals to their customers, floating the scenic riverway. On July 20 Lodge Company employee Adam Dietrich began preparing coffee and ferrying food and other supplies from the parking lot to the cook site via a wheeled cart. Such cook sites, especially since they are away from areas regularly used by high concentrations of people, are tempting attractions to bears. The Deadman's site is surrounded by vegetation, with three or four berry species to additionally attract a forager at the right time of year. The employee had bear spray in the portable kitchen, but he later reported having had no training on how or when to use it. When he saw a black bear fifteen feet away, between him and the potential safety of his vehicle, he moved toward the riverbank, where the bear followed him. The man lay down and played dead, but the animal came to sniff and paw him. It punctured his scalp, which later needed stitches. After Dietrich escaped into the Snake River, a passing rafting guide found and aided him, radioed the park for help, and beat his oar on a picnic table to scare the bear away. Park staff temporarily closed the site and set a bear trap. On July 31, upon capturing a brown-colored adult male black bear fitting Adam's reported description of his attacker, biologists euthanized the animal because of its demonstrated aggression toward humans. A follow-up report from the park recommended changes in cook-site procedures to prevent food from being left even temporarily unattended when a sole worker was preparing a meal, and to ensure that employees receive training in how to use bear spray and minimize bear attractants at the site.[18]

Six weeks later, a mother grizzly bear with two yearling cubs injured a bow hunter along Glade Creek in the Rockefeller Parkway. (The parkway's founding legislation allows hunting to continue as it did prior to the land being transferred from the U.S. Forest Service to the NPS, although since the federal government listed grizzly bears for protection under the Endangered Species Act, it has permitted no hunting of them.) A man from Cheyenne, Wyoming, hoping to hunt moose, camped along the gravel Grassy Lake Road west of the Flagg Ranch campground and lodge development. September 1, 1997, was opening day of the season, and the aspen-conifer forests and meadows along Glade Creek and other upper Snake River tributaries provide good moose habitat. After hunting for an hour or more in mid-afternoon, Greg Dolph saw a golden-colored grizzly bear coming fast for him through the willows. He screamed and climbed a dead lodgepole pine but could get only ten feet up before the bear reached him and bit his foot, pulling him out of the tree. At that point, the hunter stayed still, face down on the ground. The bear moved away and briefly returned several times, popping her jaws and biting the man's hand. Dolph had seen two cubs about half their mother's size within about thirty yards of her, and he assumed she left to check on them. After the bears moved off, the hunter walked to the Grassy Lake Road and waved down a passing car, whose occupants took him to Flagg Ranch ranger station for aid. Doctors treated Dolph for lacerations, bruising, and puncture wounds at St. John's Hospital in Jackson and released him two days later. Incident investigators found evidence of bears feeding on a freshly killed elk calf carcass in the immediate vicinity of the attack. The park took no management action toward the bear in what it considered a defensive attack of her cubs and kill.[19]

Grand Teton Park reported two minor injuries from bear encounters in 1998, and another incident occurred nearby. On June 5, a man and his daughter walking a trail near Jackson Lake Dam were followed by a brown-colored black bear that, once close enough, stood on its hind legs, put its paws on the woman's shoulders, and bit her head. Amazingly, the victim's father took two photos, one of the bear right next to his daughter and the other with the bear's paws on her shoulders. The woman then lay in a fetal position on the ground, where the bear bit her again on her shoulder and hands. When the bear ambled off, the pair made their way back to the parking lot and sought medical attention at the Jackson Lake Lodge clinic. The woman's puncture wounds and lacerations did not require stitches.[20] In a second incident, a woman hiking alone on the northwest shore of Jenny Lake encountered a cinnamon-colored black bear standing on two legs about fifteen yards away. Though the woman backed off, the bear approached, and when the hiker dropped to the ground the bear mouthed her head and bit her skull. The woman rolled down the bank and swam out into the lake, but the bear followed until another hiker scared it off by yelling. Rangers, who suspected that the bear could be the same one involved in the incident at the dam, tried but failed to catch the animal.[21]

In late August of that year, a Casper, Wyoming, man spent two days in Jackson's hospital after a grizzly bear bit him in the leg while the man was hiking around Arizona Lake. This occurred just east of the park and was accurately reported as being in the Teton Wilderness, although the news headline blared "Man Mauled by Grizzly in Teton Park."[22] The events in 1998 illustrate the two primary types of bear-human encounters that can turn dangerous. The first involves animals that have received human food and/or are used to the presence of people, as was apparent in the injuries on the path near the dam. The second type, like the incident at Arizona Lake, results from a surprise encounter with a bear. The incident at Jenny Lake is hard to categorize based on the limited information available, which suggests that it also involved a bear too used to humans, although it might have been an unpredictable close encounter.

Most bear-human encounters occur in summer and fall, when bears are actively seeking food and numerous people are enjoying the outdoors. But in 2001 Grand Teton National Park documented what demonstrates another, much less frequent, type of bear encounter—likely the first recorded bear-caused injury of a cross-country skier in North America. (I could find at least one more recent instance of a brown bear injuring a backcountry skier in Alaska, and several encounters between downhill skiers and bears in Europe.) The park reported it at the time as perhaps the first recorded grizzly bear–skier *confrontation* in America. Yet, as shared in historian-author Paul Schullery's *Yellowstone's Ski Pioneers*, a visiting journalist on an April 1902 ski trip into the first national park, in the company of two cavalry troopers and Fountain Hotel winter keeper Bill Wade, had quite a close encounter with three grizzlies.[23] A century later and forty air miles to the south, at about 10:30 p.m. under an almost full moon on the calm, clear winter night of March 7, off-duty park employee Jim "Oly" Olson was en route to the Upper Berry Creek patrol cabin, about eight miles west of Jackson Lake, when he was attacked by a grizzly bear. The bear bit the skier in the arm and thigh before departing into the night. Oly recovered from his injuries, and the park took no action toward the bear. Rangers investigating the scene found tracks suggesting that the bear deliberately sought the skier from about three hundred yards away and charged at about twenty yards.[24] This could have been a case of a wilderness bear being surprised to see a human, especially when the bear had likely just emerged from his winter den, but it could also have been more aggressive or defensive behavior.

Another early-season incident in the Rockefeller Parkway followed on April 15, 2001. An angler west of the Snake River bridge near Flagg Ranch heard a popping noise and looked up to see a grizzly bear on the riverbank. Victims of close bear encounters not uncommonly mention the animal popping their jaws. Facing the bridge and talking to the animal, the angler moved toward the road. But the bear followed and grabbed the man's sleeve, scratching him slightly as the man hit the bear on the nose and threw a rock at it. A park maintenance vehicle came along and scared the bear off. Rangers confirmed the bear species from tracks and claw marks in the snow.[25] Later that year, a participant in the elk reduction program was hunting near dusk on October 15, on School House Hill just east of Moran. While the hunter was separated from his companion, a bear later determined to be a grizzly, who had a cub-of-the-year, charged him. Although the man had bear spray, as required during the special elk season, he had no time to use it before the bear hit him, punctured his body, and displaced part of his scalp. The hunter was able to walk out of the backcountry with his partner, and since the bear seemed to be a mother defending her young, park staff took no action against her.[26]

Five years passed with no injuries despite increasingly visible bears, and biologists were concerned that the park was not on top of food storage to deter bears from developed areas. By mid-July 2006, several black bears had gotten into human food or garbage at Colter Bay. Staff reported to me that late in the evening of July 18 or in the wee hours of July 19, a bear bit a boy sleeping on a tarp in the campground. About that same time, a female bear with cubs-of-the-year got food at a campsite in Granite Canyon, tore into a tent, and pawed a camper's head. The injuries were not too serious, but that summer was a turning point for the park. If bear management hadn't been high on the agenda before, it was going to be from then on. Grizzly bear #399 had been gaining fame, appearing along park roads since spring. With three yearling cubs, the bear caused wildlife jams the following year as well, often in the Colter Bay and Jackson Lake Lodge areas. Connecting the two is an old wagon road—once the park's main road—about five miles long and then used by hikers, riders from the horse concessions located in both developed areas, and wagons heading for a Lodge Company–operated meal site on the edge of the Willow Flats, a popular elk calving area. On June 8, 2007, the grizzlies killed an elk calf not far from the Jackson Lake Lodge cabins. After a morning jogger had a close call but made no contact with the bears, the park closed an area around the lodge and north halfway to Colter Bay. Access to the cookout site continued only from Colter Bay. Standard protocol called for a five-day closure. Unfortunately, the day after rangers appropriately removed the closure signs, Dennis Van Denbos, visiting from Lander, Wyoming, walked down the wagon road behind the lodge in the early morning of June 13 and had a surprise encounter with #399 and her cubs, who had killed another elk calf. The man played dead but was bitten on the back and buttocks before two concession workers heading for the breakfast cookout frightened the bears away. The employees assisted the victim, who was treated at the hospital in Jackson. Park staff made no effort to trap bear #399 or her cubs for defending space around their kill.[27]

Even though the 2007 attack was not the first time staff had to respond to a grizzly bear–human encounter, bear #399's public visibility stimulated management action by park officials and partners. After a few weeks, the Grand Teton Lodge Company requested permission to cease offering their cookout for a while, and park managers agreed. Offering the cookout was then considered a required activity in the concessioner's contract, but it was later changed to an authorized (meaning optional) service. By summer 2008 the park had set up a regular seasonal closure of the Willow Flats south and west of the lodge. The area had essentially no trails and little public access except to the former wagon road and cookout site, which were kept open despite the adjacent closure. Added to the park's compendium of regulations was the new prohibition of public entry from May 15 through July 15 each year, to protect bear foraging opportunities in the elk calving area and to lower the risk of human-bear encounters.[28] By 2019 the concessioner had ceased using the Willow Flats cookout site, and upon completion of the next concession contract, the opportunity for that service would altogether end at that location. The Lodge Company had already removed their property, such as picnic tables, and efforts were underway to get rid of the remaining physical infrastructure, including the concrete pad and kitchen grill where meals had been cooked and served.

On October 30, 2011, a hunter taking part in the elk reduction program surprised a grizzly that may have been on a carcass. Jackson resident Timothy Hix was walking alone in a treed area along the east side of the Snake River, north of Blacktail Ponds Overlook, between the river and highway 287. Although the hunter carried bear spray, he had no time to use it before the bear charged him. The man rolled into a ball but suffered several deep bites; he was able to call for help and rangers responded.[29] No management action was taken on-site, although in response to another incident, in which a bear was killed the following year in the same area, the park and the Wyoming Game and Fish Department made some changes in the elk reduction program to reduce the risk of hunter-bear encounters.

Another bear-caused human injury occurred in Grand Teton Park on May 19, 2024, when a solitary hiker on Signal Mountain surprised a grizzly bear, thought to be a female that had a cub. The man saw a bear at close range and drew his bear spray but reportedly did not have time to pull it from its holster before he was attacked by what he assumed was a mother bear. The man fell to the ground and played dead, still holding the repellent. Fortunately, although the bear bit the man several times, the animal also bit the bear spray canister and then ran off. The hiker was able to make a cell phone call for help and received a helicopter rescue ride to the Jackson, Wyoming, hospital, where he was treated and released the following day.[30]

STAYING SAFE IN BEAR COUNTRY

Be Alert. Use sights, sounds, and smells to watch for bear activity, especially in forested and brushy areas and near streams.

Travel in a Group. Keep people together, especially kids. Leash dogs or leave them at home.

Make Noise periodically to avoid surprising a bear.

Carry Bear Spray at hand—and know how to use it.

Avoid Traveling at Night, and near dawn or dusk.

Always **Keep a Clean Camp or Picnic Area**. Pack out all food and trash.

If You See a Bear, Keep Calm and at least one hundred yards distant. **Do Not Run**. Back slowly away and talk to let the bear know of your presence.

Find more information at https://igbconline.org/be-bear-aware.

Based on how the incidents were reported, twelve (75 percent) of the black bear–related human injuries involved animals that were food conditioned, habituated to people, or a combination of both. In two other cases, people were injured trying to avoid the bear(s), and reports do not provide enough information to tell whether the animals showed aggression or tolerance of human proximity. Of the eight injuries caused by grizzly bears in the park or parkway, at least six (75 percent) resulted from a single person unexpectedly encountering the animal at close range. The bears appeared to be defending a food source, cubs, or both. (The two incidents just outside the park were both surprise encounters as well.) The winter incident appeared to be a case of an aggressive bear pursuing a lone skier, and the incident with the Snake River angler who was followed by a bear could have also resulted from early-season bears being unaccustomed to people or surprised by a person in an area with very little human use.

Given the millions of visitors that come to Grand Teton National Park and the Rockefeller Parkway each year, the number of humans injured by bear encounters is minuscule. And fortunately, most of the incidents on record were minor, even amazingly so. People were mouthed or scratched by bears, even grizzlies, who pulled on their sleeves or nipped at their legs, when clearly the animals have the power to do far more serious damage. In these parks and across bear country, people endeavor to keep such encounters to a minimum for the benefit of both species. May the record of injuries stay low.

13

Guns, Snares, and Bears

Oh, for the reported image mentioned in this "best bear story" shared by Mark Rockefeller, Laurance Rockefeller's nephew, which occurred at the former JY *Ranch in southwestern Grand Teton National Park in about 1976. Mark's father, Nelson Rockefeller, served as vice president of the United States under President Gerald Ford in the aftermath of the respective resignations of Spiro Agnew and Richard Nixon from those elected offices. During a family visit to the ranch, Mark recalled "having lunch in the dining room when a black bear walked up onto the deck and proceeded to stand up and put its paws against the glass. Nelson then went over to the window and I recall we got a picture of the two together separated only by the thin clear glass. I believe the Secret Service later put a 410 shot into the bear's butt to discourage it from further visits."*[1]

Whether the federal agents discouraged the bear from the porch or not, it was and is not uncommon for people to take up arms or other tools in defense against or in pursuit of bears in Jackson Hole and elsewhere. The archaeological record and oral histories tell us that prehistoric valley residents and visitors trapped and hunted wildlife in unknown numbers, but evidence suggests that mountain dwellers from 10,000 to 500 years ago primarily sought animals preferred by hunters today—deer, elk, bison—along with bighorn sheep, which were evidently more numerous and widespread than today. Native Americans also killed bears for meat and fur.[2] The earliest Euro-American settlers commonly hunted and trapped game, including bears, and the tradition continued throughout Wyoming, before there was any idea of a Grand Teton National Park.

With the establishment of Yellowstone, the vision for national parks was that they would not be open to hunting, and most are not, even today. Yet early conservation measures in North America often were proposed by hunters, and even today this tradition continues as numerous individuals and organizations support ethical hunting and work to save habitat for birds, fish, and wildlife. The first of a number of preserves that at one point amounted to more than 2.3 million acres of Wyoming, the Teton State Game Preserve created in 1905 was more than 500,000 acres on the Teton National Forest. Much of the land later became Grand Teton National Park or the John D. Rockefeller, Jr. Memorial Parkway. Its primary purpose was to preserve spring, summer, and fall range as well as a breeding ground for elk; domestic livestock grazing was not allowed. Hunting for elk and perhaps all species—the record is unclear but suggests there were other limits—was initially prohibited. A 1917 column in the *Philadelphia Inquirer*, one of the nation's longest-running newspapers, revealed national interest in and concern for wildlife in Wyoming and across the west. Highlighting Dr. William T. Hornaday and the Wild Life Protection Fund, of which he was a trustee, the article mentioned campaigns to protect big game species and bemoaned opening half of the Teton Game Preserve to elk hunters, calling it "desecration of a game sanctuary." The writers also complained that not one of the forty-eight states protected bears throughout the year and implied that hunting closures in some months could ensure that the "grand sport of bear hunting" would continue.[3] In 1921 or 1922 Game and Fish Commissioner Bruce Nowlin asked for and apparently received approval to prohibit the use of poison to destroy predatory animals in all national forests and Wyoming state game preserves.[4]

Eventually, hunting and trapping was permitted on much of the Teton Forest and on private land. In 1890 John Sargent homesteaded land on a heavily forested peninsula of Jackson Lake in the center of today's park, now the site of the AMK Ranch. The history of what was once several separate properties is entertainingly told in *A Tale of Dough Gods, Bear Grease, Cantaloupe, and Sucker Oil.*[5] Historic buildings that remain include a house given to and now named for former caretaker Slim Lawrence and the sprawling rock-and-log Berol Lodge. The ranch is named for owners Alfred and Madeline Berol and their son, Kenneth. Prior to the Berols' arrival, William Johnson bought the Sargent homestead in 1926, added other log structures, and extended what he called his Mae Lou Lodge to two stories because his wife was afraid of bears and would not sleep on the ground floor. It was just as well that she probably didn't know that, a few years before, a large black bear had gotten stuck trying to climb the stairs of a storage shed at Leek's Lodge, less than a quarter mile to the south.[6]

Slim Lawrence operated as a hunting guide from the AMK. His neighbor, Johnson, enjoyed spring bear hunting and made use of bait stations, where Slim put out decaying horse carcasses around Arizona Lake, just east of today's park boundary. In his home "lodge," now named for him, William Johnson hung the pelts of black and grizzly bears that he had killed. He for a time reportedly kept two pet bear cubs. Alfred Berol also hunted bear near Arizona Lake and advocated for the hunting of a limited number of bear and other game species under special licenses in restricted areas. Slim Lawrence and his second wife, Verba, were proficient wild animal trappers and hunters for food and other uses. Slim believed that the best bear pelts for rugs came from hibernating bears, so the couple searched for bear dens and hunted the animals in spring. The Lawrences' bear-bait "crib" consisted of a native log structure to cover smelly bear attractants, such as rank carcasses.[7] His traps were not unlike the "cubby" traps sometimes used by researchers, who set a wire leghold snare (with a stop to prevent the snare from tightening too much around a smaller bear or cub) underneath or along the path to the bait. Not inappropriately, the AMK is now owned by the federal government and operated, since 1976, by the University of Wyoming as the cooperative UW–National Park Service (NPS) Research Center. It has supported many scientific endeavors, including periodic bear capture and handling training for the Interagency Grizzly Bear Study Team and park staff.

Valley news coincident with the establishment of the first Grand Teton Park often mentioned the abundance of wild game and hunting opportunities in Teton County. A Teton National Forest column in a 1929 *Jackson's Hole Courier* promoted its being one of the few places in the United States where grizzly bears, one of the state's greatest assets, occurred in sufficient numbers to tempt hunters.[8] Nonresidents then paid $60 for a tag usable for elk, deer, and bear, plus $25 for the required guide license; added money was spent on the actual guide, meals, and lodging for out-of-staters.[9] The following year, nonresidents could secure a permit to kill three bears for $25 above the cost of a regular license, while residents could kill a bear at any time of year except in the game preserve.[10] In a 1931 "Game Sensus" [*sic*], the Wyoming National Forest (part of the former Yellowstone Forest Reserve and today's Shoshone and Bridger National Forests) reported the most bears in the region with a total of 327 black and brown (colored) bears. The Teton Forest came in second with 255 but had 46 of 81 "silvertip" grizzlies. The report did not separate species in noting 38 bears killed.[11]

Local newspapers in the 1930s and for decades afterward often noted the names of people who were in the valley to hunt bears, elk, or other game from spring into fall, only sometimes noting whether they succeeded. In June 1933 headlines told of a man staying at a Moran lodge who approached a game warden to buy a bear license. The warden thought the man looked familiar and asked him if he was "a reforestation army guy," while suspecting he was a nonresident looking to grab a cheaper resident tag. The visitor gave his name as Clark Gable from California. (For younger readers, he was a major movie star—the Brad Pitt or Chris Hemsworth of his day—best remembered for his lead role as Rhett Butler in the 1939 film *Gone with the Wind*.) Gable ended up buying only a fishing license and filling his limit during four guided days on Jackson Lake. He said he hoped to return to hunt bear the next year when he had more time.[12] There was no report of him visiting again.

For much of the mid-twentieth century, Wyoming authorized both spring (beginning in April and generally lasting through June) and fall bear hunting seasons. A hunter could take either species of bear but not cubs or females with cubs at their side. It appears that no bear or other game hunts were allowed in the Teton State Game Preserve until 1936. Much of the hunting then, as now, was in the Thorofare area abutting Yellowstone's southeastern corner. But bear hunts also occurred in what is now within or adjacent to Grand Teton National Park, along Ditch, Spread, and Pilgrim Creeks and around Jackson and Jenny Lakes, as well as in today's Rockefeller Parkway along the upper Snake River. From an outfitter camp run by Ben Seaton and Bennie Sheffield on Two Ocean Lake, a Chicago hunter got a black bear, in May 1940, that was "so old that its teeth were worn to the gums." His hunting partners kept after grizzlies, unsatisfied with taking only a black bear.[13] (Incidentally, Wyoming is not known for huge black bears; only half a dozen or so of some seven hundred animals listed in Boone and Crockett records come from the state, none from Teton County. Record-sized animals typically come from Pennsylvania, Virginia, and the upper midwestern states as well as Arizona, Utah, California, and Alaska.)

Not everyone favored bear hunting, especially of grizzlies. The *Wind River Mountaineer*, a weekly missive published in Lander, Wyoming, posted an opinion in 1931 titled "Sport (?)" It questioned the continued killing of bears when there were "only thirty-seven silver-tips left in Wyoming, including those in the [Yellowstone] park."[14] Yet it wasn't until 1939–1940 that the Wyoming Game and Fish Department even mentioned grizzly bears in their annual reports, which had been prepared since very early in the twentieth century. According to head warden Lester Bagley, there were then 480 grizzlies in the state but mostly in Yellowstone, and hunters had killed 18 in the two previous years. Also in 1940, the U.S. Fish and Wildlife Service shared results of their third nationwide census of big game, crediting Wyoming as being home to almost 44 percent of the nation's 1,100 grizzly bears, second only to Montana. (Alaska, not then a state, was not included.) Hunters killed 171 bears in Wyoming in 1945 and 100 in 1946, 40 on the Bridger National Forest alone.[15] Regardless of varied sentiments, hunting of both bear species continued in Jackson Hole and elsewhere in the state.

An undated historical photo shows a large grizzly labeled as the record bear for Jackson Hole, reportedly killed at Polecat Creek by a Bureau of Reclamation superintendent who was building the Grassy Lake Dam "and

FIG. 13. Grizzly bear killed by hunter, ca.1938 at Polecat Creek in what is now the Rockefeller Parkway. Jackson Hole Historical Society and Museum 2005.0163.004.

winched out by Slim Lawrence with his Dodge Power wagon."[16] The dam was built between 1937 and 1939, and Boone and Crockett Club records list a grizzly killed in Teton County by C. C. Craven in 1938, scoring 25 1/16 based on the combined measurements of the greatest width and length of the skull without the lower jaw. That bear, legally hunted in what is now the John D. Rockefeller, Jr. Memorial Parkway, was the largest bear killed in Wyoming for decades. (It was slightly surpassed in score by a grizzly shot near the head-waters of the Yellowstone River in 1960 and then by another grizzly bear skull picked up at Eagle Creek, east of Yellowstone Park, in 1961.[17] For more than half a century that animal has held the record size for a Wyoming bear.)[18]

As citizens, politicians, and the two primary federal agencies involved—the U.S. Forest Service and the NPS—weighed first the creation and then expansion of Grand Teton National Park for three decades in the mid-twentieth century, hunting or the potential lack thereof was sometimes an issue. A 1925 presidential commission had recommended against NPS control over land within the state game preserve and in favor of continued state wildlife management through regulated hunting as appropriate.[19] A 1943 news report described bears as being in the middle of the debate: Wyoming Game and Fish Commissioner Lester Bagley reminded hunters that Teton County, including part of the Jackson Hole National Monument, was open to bear hunting, having heard that Monument managers had warned the opposite. Bagley reportedly told Jackson game wardens to disregard the Park Service's stand.[20] The park's *1944 Annual Wildlife Report* included a map of the hunting zones in the preserve portions of the Monument, showing that areas north and west of Jackson Lake were open to moose, elk, mule deer, and bear hunting from September 5 to October 5.[21] By 1947, with little evident fanfare, the state replaced its game preserves with game management units, including in Jackson Hole, where much of the former preserve land was included in park expansion proposals.[22]

Though there was certainly opposition to the cessation of sport opportunities in the new park, in the end controversy over the loss of bear hunting appeared to fade. While Commissioner Bagley continued to argue strongly against park expansion, his stated concern was all about elk. *Crucible for Conservation* similarly does not mention the potential loss of big game hunting as integral to the debate, save for concern around management of the substantial Jackson elk herd (addressed in the 1950 park act by authorizing a special elk reduction when jointly determined necessary by the state and the Park Service). The broader debate over the body of law and tradition assigning "ownership" of wildlife to each state versus the NPS mandate to conserve park resources for future generations, affirmed by courts favoring the federal role in wildlife management, provided a not-inconsiderable tension that, to a degree, still exists today, influencing interagency relations. However, the new Grand Teton National Park legislation did not allow hunting, and it ceased for all species save for elk, for which only portions of the area were and are open to the reduction program.

Beyond hunting for sport or for meat, Jackson Hole residents also trapped to control animals on their properties. Slim Lawrence killed bears for causing damage or ran them off using dogs. On the morning of May 14, 1934, he shot at three grizzlies, wounding one that subsequently charged him. He fled first to the chicken coop then, as his dog, Cap, held the bear at bay, Slim climbed onto a roof to escape harm. He found the dead bear the following day, several hundred yards toward Sargent's Bay. Whether or not it was true, as Slim believed, that Yellowstone Park rangers dumped problem animals in the Arizona and Pilgrim Creek areas east of his home, the area seemed plentiful with bears—he claimed that thirty to thirty-five continually broke into a cookhouse and meat house used by workers building the Berol Lodge in the late 1930s.[23] Another event caused the demise of a presumed black bear that broke into the Elk Ranch meat house in October 1935. After the animal grabbed a quarter of beef, cowboy Howard Henrie shot the thief.[24] These incidents dropped off as private lands were acquired for addition to the park.

In 1950 Wyoming wildlife officials estimated that slightly more than two thousand black bears and ninety-one grizzly bears lived in the state.[25] The Park Service estimated that grizzlies numbered about two hundred in Yellowstone but were "very rare at Grand Teton park, Wyo."[26] A person could legally hunt both bear species and kill a grizzly bear in defense of property or human life. Outfitters continued placing ads advertising their services for pack trips, guiding, and bear hunts. Black or "brown" bear hunters were successful at Pilgrim, Pacific, and Spread Creeks east of the expanded park borders. Local newspapers still regularly reported visitors in town for bear and other game hunts: "Henry Vogt, of Chicago, hunting with Nick Dietrich, turned in the first spring bear tag to the local Game and Fish. The bear, which he shot May 19, was brown, of medium size, and had a beautiful coat."[27] It's assumed that "brown" in this case refers to a brown-colored black bear, since a September story specified that the hide of a "huge 600-lb grizzly bear shot by a hunter" on opening day of hunting season would make a wonderful rug for the Pennsylvania man.[28]

Five years later, the state halved their estimate of grizzlies outside Yellowstone. This was due to slow reproduction and hunting pressure on the small grizzly population. In 1959, when both spring and fall bear hunting was permitted and more than 75,000 grizzly bear licenses were issued, the state considered future season closures due to the animals' slow reproduction and continued hunting pressure on the small grizzly population.[29] But bears were still being killed, possibly even in the wrong places. A report on the history of grizzly bear mortalities from 1959 to 1987 lists what is described as a "legal hunter kill," near Moran Junction in 1961, of a one-year-old dependent male, despite the prohibition on killing cubs attended by mother bears. Although the report acknowledges that plotted locations may be approximate, the map places the grizzly's death near where the Snake River empties out of Jackson Lake, squarely in Grand Teton National Park.[30]

Wyoming reconsidered grizzly hunting again after the bear's inclusion in the first, less well-known endangered species list in 1966. Only seven to eight grizzlies had been taken in each year just prior to the new attention on the species. Then, after the state proposed a moratorium, hunters killed twenty-one grizzly bears in 1967, perhaps fearing it could be their last such opportunity. The state's season was paused in 1968 and 1969 but reopened in 1970, when thirty hunters received permits. Wyoming issued twenty-four grizzly permits in 1971, sixteen in 1972, and twelve each in 1973 and 1974—eight for Park County, east of Yellowstone, and four for Teton County.[31] In what did become the last season, a hunter killed a three-year-old male grizzly bear practically on the Grand Teton Park boundary with the Bridger-Teton National Forest.

At the time, it was by no means certain that grizzly bear hunting would cease. A 1974 National Academy of Sciences (NAS) report had been commissioned to evaluate the status of the ecosystem's grizzly bears in the wake of the Craigheads' less-than-amicable departure from Yellowstone Park. The authors did not suggest a cessation of all hunts, although they advised states to regulate hunter kills and hold human-caused bear deaths for the entire population below ten annually until further study demonstrated that higher mortality levels could be sustained. The strongest urging from the NAS group was for more independently led cooperative research to better understand the population's status and to provide recommendations for its management in and beyond Yellowstone.[32] Both Montana and Wyoming objected to the proposed listing of the species, arguing for continued state control over bear management, and Wyoming had already planned to suspend grizzly hunting seasons in 1975 and 1976 while more information was gathered on the population's status.[33] But it was not enough to prevent the federal government from applying Endangered Species Act protection. Grizzly bear hunting in Wyoming and most everywhere else in the lower forty-eight states stopped. Sport hunting for bears in what had become the nearly 310,000 acres of Grand Teton National Park had long since ended.

The parkway was another story.

The (Not-So-) Forgotten Parkway

One spring, several years after I'd moved south from Yellowstone to work for Grand Teton National Park and the adjoining John D. Rockefeller, Jr. Memorial Parkway, staff received a bear sighting report from a small family of "hotpotters"—folks who enjoyed natural hot springs like a hot tub. While soaking in a lovely surface-fed stream heated by runoff from Huckleberry Hot Spring, the parkway's largest thermal feature, the visitors saw a big grizzly bear watching them from the bank above, where they had left a day pack. As I recall the tale, whether exaggerated or not, the man in the group was worried enough to consider sending his partner and child to escape up the opposite bank of the stream while he, if necessary, would stay to distract the bear or, in the worst case, sacrifice his own safety on their behalf. Fortunately, no such heroics were necessary; the bear moved quietly off so the party was able to return safely to the trailhead. Not long after, I saw a photo provided by a park employee of an impressively massive light-colored boar near the Flagg Ranch development. It might have been the behemoth that had loomed over the group in the warmed creek, surely heightening the pulses of the wilderness adventurers on one unforgettable experience.

When I became chief of science and resource management for Grand Teton Park and the parkway, I assumed that, to protect the thermal features so unique to the ecosystem and the world, rules prevented swimming, bathing, or wading in hot springs themselves. I occasionally ventured off the Grassy Lake Road to Huckleberry and Polecat Hot Springs and saw a few fellow hikers—and, once, a small group of hotpotters—doing what was legal in Yellowstone, soaking in a nearby warmed creek. Since people generally changed (or temporarily shed altogether) clothes to soak and frequently had snacks or drinks, the above-mentioned incident prompted us to place a steel bear box near the hot spring where future visitors could safely stash food and gear before they entered the water. Adding even small infrastructure to the backcountry was typically a subject for considerable internal, and sometimes external, deliberations. Nonetheless, in May 2010, I accompanied a team of biologists and rangers on a bit of an adventure balancing a three-hundred-pound bear box atop a wheeled rescue-type litter along a mile of trail, across a simple plank footbridge over the stream, and up a small hill to the largest hot spring. What I recall most from that day was that only then did I discover that, unlike in the geyser-rich park to the north, Grand Teton managers had always allowed hotpotting in the actual thermal features of the parkway. Law enforcement rangers present admitted to lively exchanges with their Yellowstone counterparts who were frustrated trying to explain to visitors why the rules were different in two adjacent Park Service–administered areas. I was, frankly, flabbergasted.

Employees who preceded me said "it had always been this way." During a time of growth in the nation's recreational areas, and after signing the National Environmental Policy Act in 1970, then-president Nixon issued a challenge to agencies for cooperative projects to demonstrate his policy of "New Federalism." In response, National Park Service (NPS) director George Hartzog and U.S. Forest Service chief Edward Cliff in 1970 proposed a new parkway on existing federal acreage, and the secretaries of interior and agriculture concurred. They preferred using existing federal land to extending either Grand Teton or Yellowstone National Park. The proponents also advocated for naming it to honor John D. Rockefeller, Jr.'s numerous contributions of lands to the nation. U.S. Senator Cliff Hansen of Wyoming was the major sponsor of the bipartisan proposal, which Congress easily approved. Nixon signed the parkway into law on the Park Service's fifty-sixth birthday, August 25, 1972, and assigned management responsibility to Grand Teton, as that park had managed the roadway south of Yellowstone since at least 1950.[1] At the time, the NPS had separate (very thin, by today's standards) policy manuals for units categorized as "natural," "historic," or "recreational." Recreation areas included national lakeshores, seashores, urban parks set up in the 1960s to "bring parks to the people," and parkways, which often offered a variety of activities near the scenic roads. Guidance about resource protection was more lenient than in the older designated national parks. The 23,000 acres of Rockefeller Parkway transferred from the U.S. Forest Service, which typically permits more "multiple uses" than parks, included a sixty-seven-acre commercial Huckleberry Hot Springs Resort—including a gas station, camping and picnic areas, and a sizable warm-water swimming pool with a slide. Those facilities operated before and after the NPS took over, until they were closed due to fiscal and safety issues and the site was largely restored in the 1980s.

Park managers had apparently never considered whether continued soaking in the outdoor thermal features sans resort was appropriate under the substantially updated NPS Management Policies (prompted by several legislative and judicial actions in the 1970s) that have since applied to *all* park units. They define appropriate recreational uses as those that are appropriate to the park's legislated purpose and that do not cause unacceptable impacts to the resources or create an unsafe environment for visitors.[2] The NPS recognizes that most uses affect park resources; the standard is that the integrity of resources must be left *unimpaired*, which requires park managers to make judgments about the severity, timing, and duration of uses and related impacts. Fans of warming themselves in the parkway's hot springs can blame me for pitching closure of hotpotting in all thermal features, out of concern for human safety and for the ways that human use disturbs the springs' sediments, water chemistry, algae, and delicate sinter edges. As I interpreted the agency's mission, even the parkway's unspectacular thermal features, though they pale in number, size, and heat level in comparison to Yellowstone's thousands of springs, should be conserved "in such manner and by such means as will leave them unimpaired for the enjoyment of future generations."[3] I came

FIG. 14. Rockefeller Parkway, looking north toward Yellowstone, Snake River on left, road on right. National Park Service photo.

as close to using the word "impairment" to describe use of thermal features as I had at any time in my forty-year NPS career. And in 2014, in what I credit as one of his major accomplishments, new superintendent David Vela approved a change in parkway regulations, to prevent human entry into "waters solely of thermal origin," matching Yellowstone's protective language. Visitors may still enjoy a soak in the nearby streams warmed by runoff from the hot pools.

My obvious digression from bears is meant to illustrate how the Rockefeller Parkway has arguably been and perhaps remains one of the somewhat forgotten, or at least overlooked, units of the National Park System.

The very name connotes "just passing through" on the road between the world's first national park and the slightly less renowned smaller and younger "grand" park to the south. For most of my thirty-plus-year tenure in the area, superintendents and other staff like me introduced themselves as assigned to either Yellowstone or Grand Teton Park, not mentioning the connecting corridor. Literature and other communications typically list two parks, three states, two national wildlife refuges, and several national forests (the latter also numerically challenged by the bureaucratic combining of legislatively distinct forests into administrative units such as the Caribou-Targhee and the Bridger-Teton) in the Greater Yellowstone Ecosystem. The parkway was not included on official government letterhead stationery used by Grand Teton National Park until 2012, the fortieth anniversary year of the unit's establishment. During and since that anniversary, employees started to be more inclusive in describing themselves, in professional introductions and on their business cards, as working for both the park and parkway. Really, except for bureaucratic or legal nitpickers, what does it matter?

Just as the hot springs were arguably under-conserved for decades of the parkway's existence, a few other things got overlooked in the landscape linking the two most famous Wyoming parks. Although Congress called for agencies to assess areas of sufficient size and character for potential wilderness designation beginning in the early 1970s, by the time the U.S. Forest Service embarked on its ambitious "RARE II" roadless area evaluation, the Bridger-Teton Forest no longer administered the lands. Not until 2013 did the NPS consider this possibility and determine that 91 percent of the unit met wilderness criteria. The parkway does receive a small distinct line-item allocation (only for interpretation, maintenance, and law enforcement) in the NPS's budget "green book," although no one closely tracks how much money or staff time is spent in each locale. The lack of a resource management budget for the parkway has not prevented staff from monitoring and protecting wildlife there, although the unit deserves some designated funding for natural and cultural resources. Typically, staff assignments wax and wane in parks across the nation as workers cannot spread their time and attention equally across all acres of any unit. Outlying districts distant from headquarters and major visitor destinations, in perception or reality, can get short shrift in public attention or staff presence. By the time I retired, rangers seldom lived or worked out of the Flagg Ranch developed area; managers often preferred staff to be stationed in busier parts of the park and many employees enjoyed being closer to town. The ranger station had long since been unmaintained and boarded up for health and safety reasons. Staff or volunteers periodically drove up north from Colter Bay to operate a small temporary visitor contact station only in summer, if at all. Managers argued such measures were justifiable due to the relative lack of use the parkway receives. Still, there's a danger of being "out of sight, out of mind," and the absence of visible Park Service presence may affect the recognition and protection of the parkway's visitors and its resources.

When it comes to wildlife management, the parkway differs from the two parks between whose crusts it is sandwiched. Yellowstone, predating the three states that overlap its boundaries, has exclusive federal jurisdiction. State agents do not enter to enforce laws, which of course do not permit hunting. Grand Teton and the parkway, like many younger units, have concurrent legal jurisdiction: state and federal agents cooperate to enforce laws such as regulations on fishing, which parks typically permit. As previously mentioned, Grand Teton's legislation does not allow hunting, except for the unusual and explicitly named elk reduction program, for which both park rangers and Wyoming Game and Fish officers enforce the associated regulations. The parkway's own legislation permits both fishing *and* hunting, which were grandfathered in when Forest Service lands were converted to the National Park System. The mostly forested unit has geology and vegetation like that in southern Yellowstone—volcanic soils, lodgepole pines, and remnants of large periodic wildfires, such as those that occurred in 1988 and 2016. Tall willows line the banks of the upper Snake, designated as a wild river in the parkway. There is light use along the dirt-and-gravel Grassy Lake Road and few trails. It is a wildlife-rich area.

Black bear hunting thus occurred before parkway designation and has continued since. The area is part of a large black bear hunting district, Conant Basin/Area 21, making it difficult to figure out how many bears are taken in the parkway itself. In the 1980s and 1990s, the U.S. Forest Service was embroiled in a multiyear controversy over allowing legal hunters to use bear-bait stations on its lands. The practice, though it stayed below the radar in the parkway, had long been described by critics, even some that support hunting, as "not sporting." The controversy went national by 2003, when Wyoming opposed a measure proposed in the U.S. Congress to prohibit bear baiting on federal land across the nation, although by the late 1990s it was already banned in Forest Service wilderness areas and most wildlife refuges. It had also stopped in the Greater Yellowstone Grizzly Bear Recovery Zone, which includes the entire Rockefeller Parkway. Area 21 hunters without using baits killed seventeen black bears, eleven in the spring season and seven in the fall of 2021, the last year for which reports were available as of this writing.[4] In a bit of public confusion, the hunt area map showed the parkway, along with Yellowstone and Grand Teton, as closed to hunting, although the written description accurately included it as an open area. Ever since proposals to remove grizzly bears from the list of threatened species have resurfaced, there has been growing focus on the potential to hunt them in the connecting corridor between Yellowstone and Grand Teton National Parks—for good reason, given the area's history.

An interagency team documented at least seven grizzlies killed between 1959 and 1987 within what became the Rockefeller Parkway. For several other deaths, the precise locations are a challenge to confirm as reported place names or locations suggest they could have been in or out of the boundary. Young males were legally taken at Polecat Creek in 1962, at Dime Creek in 1964, and near Flagg Ranch in 1971. Bears of unknown sex and age were hunted just south of Tanager Lake near the Yellowstone Park border and at Glade Creek west of the Snake River in 1965 and 1967. A park ranger—likely from Yellowstone—dispatched a grizzly bear found wounded along the Grassy Lake Road in 1970. In late September of 1972, an adult female was illegally killed near the old Flagg Ranch cabins that were located on the north bank of the river just off the main road joining the two parks. In 1976, the first year in which the NPS was expected to report grizzly bear mortalities in the parkway, a black bear hunter said he mistakenly killed an eight-year-old grizzly in Huckleberry Flats, near the hot springs of Polecat Creek, mentioned in the chapter's opening story.[5]

By the mid-1980s, managers across the ecosystem were zeroing in on an increase in real or potential bear conflicts with humans. Although no legal hunting of greater Yellowstone grizzlies had been allowed since their listing, there was simmering debate, at least among wildlife managers, over who should be able to kill "nuisance" bears. In late 1989 the U.S. Fish and Wildlife

Service, then headed by Jackson Hole native John Turner, proposed a revision in regulations that would permit "specially authorized persons to take dangerous or incorrigible" grizzlies to minimize human injuries and conflicts.[6] The announcement of an open public comment period received widespread media attention. Then–recovery coordinator Chris Servheen was reported to have proposed the change to provide flexibility and to allow sportsmen to participate in grizzly bear management, although the three states in the ecosystem had reportedly requested the action.[7] News items mentioned that hunting, if approved, would not occur in Yellowstone and Grand Teton National Parks. However, the preliminary Federal Register notice of the proposal did not mention the Rockefeller Parkway—and *did* include it in the mapped area considered for nuisance bear hunting.

Public comment on the idea was minimal. Of seventy-four letters sent to Servheen, 71 percent of respondents opposed and 29 percent favored the nuisance hunt. A third of the commenters included remarks such as "you guys are crazy" or "you are a bad person."[8] The director of the Wyoming Game and Fish Department penned a letter to the editor defending the federal recovery coordinator against unwarranted personal attacks.[9] Public opposition included a tongue-in-cheek call by the environmental group Earth First for a protest hunt of "nuisance bureaucrats."[10] I found no record indicating that the NPS commented on the proposal or expressed concern about grizzly hunting in the parkway. After several years of agency indecision, the nuisance hunt idea appeared to quietly slip from the public eye. This was a combined result of northwestern Montana, in 1991, losing their long-standing limited grizzly bear hunt; new concerns over grizzlies being sent to zoos or to the newly planned Grizzly Discovery Center in West Yellowstone, Montana; and the U.S. Fish and Wildlife Service moving to complete a new Grizzly Bear Recovery Plan in 1993, which reportedly would not include any change in how agencies could manage nuisance animals.[11]

Meanwhile, the carnivore's presence was increasing; there were ninety-six grizzly bear observations in the parkway from 1976 to 1992. Eleven radio-collared animals were responsible for two-thirds of the sightings, and at least five adult male and three adult female grizzlies were thought to be in the area.

Staff were doggedly addressing bear attractants. Flagg Ranch and the Grassy Lake Road were among the highest priority areas to receive improved trash and food storage containers. Many were paid for by the concessioner operating the lodge and campground. Botanists mapped vegetation in the parkway for an intensive but later-abandoned effort to produce a model of cumulative effects on the ecosystem's grizzly bear habitat. Bear researchers and managers increasingly recognized its value and the need to protect it, both from loss or fragmentation and from expanding levels and seasons of human use.

The Park Service approved a plan in August 1993 to update the Flagg Ranch commercial development by replacing forty-two old cabins and the store complex with a newly built lodge several miles from Yellowstone's south entrance, outside the floodplain of the Snake River. The new 21,000-square-foot building would have more than one hundred rooms, a hundred-seat restaurant, and a gift shop ready to open in July 1995.[12] At that time, Flagg Ranch was still a significant launching-off point for winter snowmobiles and snowcoaches heading into Yellowstone. (Ironically, by 2014, after years of controversy surrounding snow machine use, the larger park severely limited the numbers permitted through each entrance. Winter use of the Rockefeller Parkway has been quite low since, though the area remains popular with lodgers, anglers, and small numbers of river users in the summer and fall.) When assessing environmental impacts of major decisions, agencies must consult with the Fish and Wildlife Service on how proposals may affect listed species. Such consultations under Section 7 of the Endangered Species Act result in biological opinions and, typically, an agreement on necessary measures to lessen harmful effects, in this case on grizzlies. Mitigations for the Flagg Ranch plan included restricting the seasonal dates, daily hours, and numbers of commercial river trips and horse rides; limits on new trails; converting the Park Service–operated Snake River Campground (which had been closed since the 1988 wildfires) to a day-use picnic area; and replacing an undefined garbage dump with an indoor trash storage and loading area. Due to the expected net loss of 632 acres of summer habitat for bears, the Park Service promised to close the Grassy Lake Road to public motorized use from April 1 through May 31 each year. Biologists predicted that this would effectively provide a gain of 761 acres of undisturbed spring habitat.[13]

The recovery plan recognized that securing sufficient, effective habitat was crucial to maintain the bear population. Scientists had long been assessing vegetation conditions, along with the effects of human use and access provided by roads and trails, on the animals. In 1998 the Yellowstone Ecosystem Subcommittee of managers, after considerable debate, made the major commitment to maintain "no net loss" of grizzly bear habitat in the recovery zone. Since then, development of roads and other infrastructure has been limited to existing areas or has required an equal amount of acreage to be restored nearby if habitat is lost due to construction. While this policy affects less than half of Grand Teton National Park, it preserves habitat in the entire Rockefeller Parkway, as it does in Yellowstone, helping recovery and subsequent maintenance of ursids.

A synopsis of efforts to maintain grizzly bears comes from information widely available on U.S. Fish and Wildlife and other websites.[14] Although the species' listing came in 1975, recovery planning lagged until the first such document was completed in 1982. When updated in 1993, the plan called for an added conservation strategy to define actions that would ensure maintenance of the bear population after it was removed from the act's protection. The initial years-long effort that I and others worked on in the 1990s resulted in the original such strategy, which was approved in 2002. After greater Yellowstone grizzlies reached the outlined biological objectives in 1998 and sustained them, the Fish and Wildlife Service proposed delisting in 2005. It again updated the recovery plan the following year and removed the population from threatened status in 2007. Following several lawsuits, a federal district judge overruled that decision based on concerns about enforcement of conservation strategy mechanisms and particularly over the failure to assess the impacts of declining whitebark pine, a key grizzly bear food, on the population. On appeal by the federal government, the Ninth Circuit Court found that the strategy did sufficiently outline adequate regulatory mechanisms to protect the bears but agreed with conservation groups concerned about the whitebark-bear nexus. That verdict prompted the comprehensive study of bear diets, mentioned in chapter 8, which identified at least 266 foods used by the ecosystem's grizzlies.

But work to remove the animal from its special legal status has continued. Readers may wonder why, as some critics put it, there is a "push" to delist grizzly bears. One reason is that it is the goal of the act to improve the likelihood that any listed species will survive, literally to "recover." Another is that there are constituents—not just those who aspire to hunt grizzlies again in the lower forty-eight states—who believe that the money and time paid by state and federal wildlife managers can be reduced at some point to permit more attention to other species or issues in need. As of March 2024 the U.S. government had listed 1,668 domestic and 699 more "foreign" animal and plant species as threatened or endangered.[15] Only fifty-four—slightly more than 2 percent—of the total, including several populations of humpback whales, the American alligator, and the Louisiana black bear, had been declared recovered, and sadly by then the U.S. Fish and Wildlife Service had declared that twenty-one species were extinct. While most of the listed populations remain and at least fifty-six improved from endangered to threatened, the low success rate has prompted periodic debates on the merits of having an Endangered Species Act. Thus, defenders have been eager to celebrate populations that have achieved recovery goals, as many believe greater Yellowstone grizzlies have done. The official population estimate was 965 as of 2023, more than triple the estimated 136 to 300 bears in the ecosystem at the time of listing in 1975,[16] and many more bears were living in a much larger area than they did in 1976. Three, then four, and nearly five decades after listing the grizzly bear, with high public and agency engagement in managing animals, humans, and habitat across state lines, could hardly be described as a rushed time frame for removing the species from the added legal protection of the act. Especially in the western United States, state Game and Fish agencies and many of their constituents contend that "home rule" is traditional and preferable to continued federal control over wildlife management. The states in greater Yellowstone have certainly been eager to regain lead responsibility for grizzly bears outside parks. Yet both the federal and state agencies need to address public concerns. The acceptance of grizzly bear recovery and delisting in the ecosystem will depend on public trust and confidence in the research and management done in the past, present, and future.

With regard to the Rockefeller Parkway, one contentious issue has dominated the debate over the bears' removal from the threatened species list. During my tenure, which overlapped with three Wyoming Game and Fish Department directors, Grand Teton officials acknowledged that the parkway's founding legislation in 1972 allows for hunting, which could arguably someday include grizzlies (and wolves, removed from Endangered Species Act protection in 2012.) However, that legislative record suggests that the predators were not considered in the provision. Rather, lawmakers were concerned about managing and retaining opportunities to hunt elk. Wolves had been eliminated from the Northern Rockies for half a century then, and grizzly bears were rarely seen and not thought to live in the parkway. But as things changed, twenty-first-century superintendents and wildlife managers consistently, if quietly, made known their concerns. The Park Service did not comment on potential bear hunting outside parks. But with proposals to delist first bears and then wolves, Grand Teton was joined by Yellowstone in expressing apprehension over visitor experiences and, since there is no hunting in most National Park System units, anticipating public outcry should hunters kill these predators on lands under their agency's jurisdiction. Managers worried, too, that hunters might accidentally shoot animals in either national park, whose boundaries with the parkway are not easily distinguished, or that animals shot legally outside could flee, wounded, into one of the parks and die where hunters could not retrieve a carcass. Both pro- and anti-hunting advocates have been vocal, even passionate, in expressing their views.

An amazing amount of highly ranked bureaucrats' time, increasingly after 2012, was spent sharing issues and (presumably) actively listening to concerns about management options for hunting the predators, once delisted. By 2016 NPS concerns had become public, as outlined in their comments on the latest proposed delisting. This rule addressed the issues mentioned above and spoke to the topic of bears as a visitor attraction and an economic benefit. The Park Service specifically asked for several things, including: 1) that once grizzly bears were declared recovered and hunting might resume, states would focus harvests away from park boundaries in more distant areas with prevalent bear-human conflicts; 2) that despite not permitting hunting in the parks, their managers be participants in annual meetings with the states of Wyoming, Montana, and Idaho to allocate "discretionary mortality" (the number, if any, of bears that could be legally killed [mostly hunted] after accounting for legal and necessary management removals of bears); and 3) for no hunting in the Rockefeller Parkway.[17] Wyoming reassured the agency and the public that there was, at that time, no intent to hunt grizzlies in the parkway or on private land inholdings inside Grand Teton National Park, but they were consistently reluctant to commit to that in writing. This is understandable from the perspective of keeping all of their legal options open in the long run.[18]

In June 2017 Secretary of the Interior Ryan Zinke announced that, once again, the federal government would remove the greater Yellowstone grizzly bear population from the list of threatened wildlife. By 2018 the three states had completed plans to reinstitute limited bear hunts, although Montana deferred proposing an immediate season. The states agreed to meet annually with representatives from both national parks and review data from the Grizzly Bear Study Team to ensure that hunting quotas do not exceed the number of mortalities that scientists believe the population can sustain. Before hunting season could begin on September 1, legal action stopped the two states, and the ruling district judge again returned the population to listed status. The U.S. Fish and Wildlife Service appealed the decision in 2019, but the relisting was upheld in July 2020.[19] Most of the agency players involved in the heartfelt discussions I took part in had by then, like me, moved on or retired. I have long likened the semiannual ecosystem bear managers' meetings to a bureaucratic soap opera: Observers can go away for months or years and return to find that the faces may have changed but they can pick up the plot line right away, which provides me both frustration and comfort.

By 2023 the federal government had agreed for a third time to formally consider whether the ecosystem's grizzly bear population warranted removal from Endangered Species Act protection. State leaders remained frustrated with the lengthy processes and threatened to sue. Wyoming congressional representatives, along with their counterparts in Montana and Idaho, had proposed legislation to remove bears from the threatened species list and to prohibit further legal challenges in 2021 and again in 2023. Yellowstone ecosystem managers finally approved updates to the 2016 Conservation Strategy. To help ensure genetic connectivity between the isolated southern bears and more northerly populations, Montana and Wyoming wildlife biologists collaborated to relocate two grizzlies, a subadult male and a subadult female, from the Northern Continental Divide Ecosystem in northwestern Montana into the Greater Yellowstone Ecosystem in August 2024. The U.S. Fish and Wildlife Service announced that they would decide on delisting grizzlies in both of those areas and throughout the lower forty-eight states in January 2025.[20]

Areas outside Yellowstone National Park were historically the last best places for hunters seeking *Ursus arctos horribilis* in the ecosystem and will be again when the Endangered Species Act no longer applies to grizzlies, doubtless prompting continued civic debates. Previously stated concerns are on record and are likely to remain in the minds of numerous organizations, individuals, and managers. Despite the relatively low attention often paid to the smallest national park unit in the ecosystem, I hope that interested citizens will not overlook this key link between its two more famous neighbors and will remain engaged on bear management issues and influences in the never-to-be forgotten John D. Rockefeller, Jr. Memorial Parkway.

Science and Teton Bears

One of the people I spoke to in compiling this history was longtime colleague and friend Dr. Henry (Hank) Harlow, former professor of zoology and phys-iology and now emeritus professor at the University of Wyoming (UW). I reached out to Hank because of his expertise studying hibernation physiology and other aspects of bears (and other species) around the world, as well as his twenty-two-year tenure as director of the UW-NPS Research Center in the park, located at the historic AMK Ranch. Hank is a lanky, effusive fellow with a full beard and intelligent humor. When I sat down to talk with him, I was not too surprised that he'd done his homework before our visit, preparing notes of incidents and sightings with dates and places at which he had experienced bears in Grand Teton National Park and the Rockefeller Parkway. Hank categorized his observations as demonstrating curiosity, habituation, or even potentially predatory—or at least dominant to threatening—behavior, the last of which was my favorite:

In about 2009 Hank canoed a calm, meandering stretch of the upper Snake River from the former Flagg Ranch site to Lizard Creek Campground on Jack-son Lake, a route on which I had previously accompanied him and another UW researcher studying beaver. This time, Hank's wife, Mary Ann, went along. Downstream a half mile from their launch spot they beached, and about twenty feet from the river they were on their knees busily completing vegetation transects when Hank stood to stretch and saw what he judged to be a subadult grizzly bear on the bluff above the water. The bear lost its footing, fell into the river, and floated out of sight. A bit later he saw the bear again—close—moving its head side to side and locking eyes with Hank, obviously not happy with the humans' proximity. In Hank's telling, the party responded by moving to stand together, with arms linked like the Three Musketeers, as he unclipped his bear spray and faced the animal. When it came within ten or fifteen feet, Hank "hollered really loud" and was relieved when the bear noiselessly moved off into nearby alders. Though Hank recalled thinking they needed to get out of there quickly, before the bear could come charging back out of the vegetation, the researchers finished clipping their plant samples before they climbed into their canoe and moved downstream. I called him out on his "typical scientist!" behavior, but he shrugged and said if they hadn't finished the work then, they'd have had to return and risk another day for their data.[1] It was perfectly logical to him.

The grizzly bear population in and around Yellowstone is the most studied large carnivore population in the world, with essentially continual research since 1959. The first three decades of that work did not occur in Grand Teton National Park. Ironically, twins Frank and John Craighead, who began grizzly bear research in Yellowstone, kept homes inside the younger park, and their families still do as of this writing. When they began their groundbreaking work, there would have been few grizzly bears to study close to home. Yet the first use of the term "Greater Yellowstone Ecosystem" is attributed to the Craigheads. They developed and deployed the first radio collars to track the large bears across the landscape in the 1960s and found that the bruins moved across Yellowstone Park's borders—even into Jackson Hole. *The Grizzly Bears of Yellowstone* (1995) presents data and other findings from their work, including maps that show known and possible den sites in what is now the parkway and bear mortalities both there and in Grand Teton Park, but these animals roamed from their capture sites in central Yellowstone.[2] Grizzly bears were popularized by the Craigheads' resulting media coverage, and of course concern over the species' management and population status. After their work in Yellowstone ceased, controversy and the need for continued studies of the population prompted creation of the Interagency Grizzly Bear Study Team (IGBST) in 1974. It is still the entity studying the ecosystem's grizzlies and provides annual reports to managers on the status of the population compared to recovery criteria.

A decade after well-documented disagreements with Yellowstone National Park officials over the speed of dump closures and grizzly bear numbers, the Craigheads engaged in proposals to consider supplemental feeding of bears there out of concern that the animals lacked sufficient food safely away from people. (No such feeding happened, due to the Park Service and other scientists doubting its necessity.) The world's first national park not only dominates public and media attention in the ecosystem; it is a lightning rod that attracts more researchers and scrutiny of scientific studies than do the surrounding land management units. Grand Teton National Park and the parkway historically are absent the drama and contention that surround Yellowstone science, especially involving bears, and I perceived that resource specialists before and during my tenure sometimes resented being eclipsed. On the other hand, what manager in his or her right mind would regret the relative lack of controversy over science in their park? Not I nor, for the most part, staff who, once they pondered it, appreciated being able to quietly conduct their research and monitoring without distraction and in collegial cooperation with sister agencies and universities.

In contrast to the early years of the Grizzly Bear Study Team, staff from both parks have been active partners in that research since the 1990s. Teton Park biologists have played a strong part in efforts to monitor and collect disease-resistant cones and support experimental planting of whitebark pine, listed as threatened in 2022, to maintain the species for its own value and as a key bear food. And, especially in recent decades, there have been fine studies of bears, bear habitat, and bears' relationship to humans and their activities outside as well as within Yellowstone. Overall, there has been less research of most subjects in Grand Teton or the parkway, but the southern parks have hosted some unique science projects that added valuable information on bears.

Wildlife research often focuses on basic ecological studies of species— better understanding where they live, what they eat, and how they spend their time. Efforts focused on grizzlies have yielded sufficient information on which to base bear management and human education efforts to benefit both ursids. Thus, there is a paucity of science specific to black bears in the ecosystem, but Leslie Frattaroli's graduate study provided valuable information on black bear diets, movements, and patterns of use in southern Jackson Hole. (The dietary information presented in chapter 8 relies heavily on her work.) For example, she found that black bears were most active in July despite its being one of the months in which human use is highest. Both male and female study subjects typically rested from at least 9:00 p.m. to 2:00 a.m. then, as expected of diurnal creatures, were active all day with peaks in the mornings and evenings. Times of peak activity, both seasonally and during an average day, appeared to maximize bears' foraging efficiency, which was high when plants such as berries were abundant and when daytime temperatures were lower. Frattaroli not unexpectedly found considerable overlap among bears' home ranges. Males ranged over twenty-five square miles, three times that of females, and more than half crossed onto land outside the park. Such information can be useful to managers of federal, state, and private land.

Another primary focus of wildlife studies is how species interact with each other. Grand Teton National Park, in contrast to Yellowstone, offered, until the mid-1990s, a site where researchers could study a protected area lacking two of its largest native predators (grizzly bears and wolves). Then those species were, at nearly the same time, reoccupying Jackson Hole such that by the middle of the first decade of the 2000s both were well established as residents of the park and adjacent areas. Scientists salivate for such opportunities, and some came eagerly to do research. Joel Berger and various associates were interested in the distribution and timing of births by female moose and in their response to grizzlies after the bears' absence from the Tetons for decades.[3] Moose calves were born to marked mothers within a nine-day period around May 27, regardless of bear distribution. To test prey vigilance toward predators, the investigators, with determined zeal such as that displayed in the opening story, placed snowballs dipped in bear feces near adult female moose that were foraging. Observers saw responses they termed "curious," evidenced by moose moving their ears forward, and "aggressiveness," when moose retracted their ears. Cow moose that lost their calves showed a delayed response to the smell of grizzlies. As bear density increased, mother moose narrowed the distance between birth sites and paved roads, apparently seeking protection from predators for their young.[4] Researchers characterized what they termed "predator-naïve" Teton moose as being noticeably lacking in astuteness. They suggested it might be too soon to assess the ecological effects of restored grizzlies and wolves, compared to predator-moose interactions in interior Alaska—where prey showed much stronger reactions, described as "behaving with resilience."[5] The scientists did offer that moose in the Tetons could readapt as they gained savviness about grizzly bears, noting that restoration of ecological processes takes time. Grand Teton Park offers a chance for future research to compare added changes witnessed in the ensuing twenty years or after even more time has elapsed.

Several projects studied bear-bear interactions and habitat overlap. Greg Holm's 1998 master's study looked at interactions between black and grizzly bears co-occupying livestock grazing allotments in the Blackrock–Spread Creek area, 15 percent of which was inside the park and 85 percent on adjacent Forest Service land. As is typical, black bears were most active in the daytime and grizzly males at night, while females were crepuscular, moving most near dusk and dawn. The two species had substantial overlap, though much-larger-than-elsewhere black bear home ranges might have been due to them moving around more to avoid grizzlies. Since the larger bears were still reestablishing themselves in the Buffalo Valley area during the mid-1990s, Holm and his collaborators suggested that black bear numbers there would decline as grizzlies increased—another yet-to-be-tested supposition.[6]

In the early-twenty-first-century comparison of black and grizzly bear diets mentioned earlier, researchers found, as expected, that grizzlies, especially larger ones, were better at obtaining protein from ungulates through predation, scavenging, or both. Even when living in areas unoccupied by grizzlies, black bears did not eat as much vertebrate meat and took in only about half as much animal protein. But the two species got similar percentages of energy from nuts, although there was a difference in the method of gathering. Grizzly bears obtained nearly all their nut energy by digging up those cached by red squirrels. Black bears typically got half their nuts by digging and only slightly less by climbing trees to reach the cones. Those who lived where grizzlies were absent got 100 percent of their nuts from trees. These behavioral differences can be at least partly attributed to black bears' better ability in and affinity for tree-climbing compared to heavier, longer-clawed grizzlies.[7] Over time, new dietary work may show changes that result from how changing climatic conditions or other factors affect bear food availability.

Managers particularly desire information on animals' interactions with humans. Beyond documenting what bears ate, Frattaroli looked at how human use affected black bears' activity patterns. During her field work in 2005–2006, nearly 1.5 million vehicles drove the Teton Park Road from Moose to Jenny and Jackson Lakes from June to August, and visitor use on trails added to pressure from residents living in or adjacent to the park near the Jackson Hole Mountain Resort. Her hypothesis that black bears would be less active when near trails, roads, and human developments did not hold true, though the researcher emphasized that it was still possible that humans were affecting bears. Since the forested, low elevations she studied offered a range of high-quality foods and provided travel corridors for black bears, it may have been worth it for the animals to tolerate people, especially under cover of trees. Educating visitors and residents about bear presence and associated risk in and outside the park has been important. The year following her field work, state wildlife managers destroyed three of the marked bears for getting into human foods or garbage outside the park, which one might view as sadly predictable.[8]

Another study looked at how recreation on a new multi-use pathway affected black bears (at the time, there were no grizzly bears) in southern Grand Teton National Park. Prompted by high demand for safe off-road bicycle paths after two riders were killed, the park obtained funds to build an eight-mile path between headquarters at Moose and the popular Jenny Lake area. Researchers gathered pre-pathway data along with more information collected during construction in 2008 and during the first three years of use in 2009–2011. Once the path was built, peak human use occurred in July and August and between 11:00 a.m. and 4:00 p.m. each day, during which trail counters detected an average high of twenty-five to thirty or more users per hour. In response, black bears changed neither their home ranges nor how often they crossed the paved trail. They did make behavioral shifts to enable continued use of habitat near the path for feeding and moving while reducing their chance of human encounters. They spent less time near the trail in midday and made more use of steeper slopes near the path compared to pre-construction. Bears rarely crossed the pathway, but when they did it was typically early and late in the day, when pathway use was low, and in spots where vegetation grew close, which might pose a higher risk of human encounters.[9] I can relate; early one morning before work my biking partner and I topped a small hill and came perilously close to a black bear grazing near the path. It was unclear which of us was most surprised as we all veered away from each other and danger. Park managers, influenced by the research and other information, closed the pathway to use from sundown to sunrise.

The park also engaged the Grizzly Bear Study Team to evaluate how the elk reduction affects grizzly bears. Other studies have shown that fall elk hunting increases the likelihood that hunters will encounter bears seeking the bounty provided by animal gut piles and carcasses. The reduction program typically occurs from late October to early December, and from 2008 to 2017 hunters killed an average of 242 elk each season, all cows (females) or spikes (yearling bulls with a single spiked antler); bull hunting was not allowed. After back-to-back years with grizzly-human encounters—in 2011, when a bear defending a carcass injured an elk hunter, and in 2012, when a hunter killed a grizzly in self-defense—the Park Service and the Wyoming Game and Fish Department revised requirements for the specially permitted program participants. They already had to carry bear spray in the park and could not use elk calls to draw in animals; new restrictions limited hunt areas in the Snake River riparian zone and the amount of ammunition that each shooter could carry. The park added extra education about safety, and rangers or researchers made personal contact with most hunters, who were quite cooperative in sharing information about their movements and observations. In 2014–2015, scientists used remote cameras and collected hair samples to extract DNA and find bears using the eastern part of the park and the adjacent Bridger-Teton National Forest. Although they hypothesized that hunting activity would temporarily attract more grizzly bears to the study area and cause heightened human risk, findings did not support that idea. Instead, bears already living in the area took advantage of the added food source. Though elk carcasses are a boon to bears in their hyperphagic run-up to denning, the lateness of the season could have meant that few bears were still awake to take advantage.[10]

Fortunately, no serious bear-human confrontations occurred between 2012 and 2023, though risk remains as in any area co-occupied by people, bears, and attractants. When Grand Teton National Park revitalized its bear management program in 2007–2008, staff made a major effort to revamp informational signs and messages intended to inform visitors of bears' presence and how they could reduce the chance of encounters. Two years later, in one of the first bear-related social science studies in the ecosystem, the park arranged for Patricia Taylor and Nanette Nelson of the Wyoming Survey and Analysis Center to evaluate how the new efforts were working. Based on their 2010 survey of 634 summer visitors to front-country campgrounds and picnic areas, the visibility or "catchiness" of signs varied. More than three-fourths of visitors noticed colorful yellow cards stapled to picnic tables and "Be Bear Aware" signs on trash cans, on bear boxes, and along park roadways. Signs warning of bears frequenting an area and saying "Danger— Trail Closed" were least noticed, possibly because fewer visitors ventured to areas where those were placed. People gained information from personal contact while registering for campsites, at park entrance stations, and from other employees—even boat captains—more than from websites. They also relied on their own experience and talking with friends or family members about bear safety. With the increasing popularity of social media in recent years, future research could investigate whether visitors' information sources change. Picnickers were less likely than campers to receive information from anywhere. Still, awareness of basic safety behaviors such as storing food, garbage, and even cooking gear and toiletries was quite high. Overall, sampled visitors recognized that grizzly bears did live in Grand Teton National Park, that black bears could be a threat to humans, that animals are unpredictable, and that running from them could cause an aggressive response. Results encouraged the park to continue its educational efforts although, as the social scientists wrote in their summary, there will always be visitors like the one who said, "I want a bear to come into the campground."[11]

Ariel Blotkamp earned her master's degree on day hikers' knowledge and fears of bears, surveying visitors in 2010. Based on 350 responses at four park trailheads, hikers from ages eight to eighty-one from forty-two states and eight countries—more than half on their first visit to the park—could have benefited from more knowledge of both bear and recommended conflict-avoidance behaviors. People hiked an average of less than four hours, and 7 percent of them saw a bear while doing so. Three primary misconceptions could affect hikers' safety: First, nearly 80 percent of respondents thought carrying bear bells would effectively alert the animals to the hikers' presence, although bear managers disagree since terrain, wind, water, and other noises often prevent bell sounds from carrying. Second, almost two-thirds of those surveyed thought hiking with food increased their danger of encounters, although there is no evidence to support that fear. Third, more than 60 percent were unaware that a bear standing up on its hind legs does not necessarily signal an impending attack but is a means for the animal to better use its eyes and nose to sense potential danger. The researcher found little relationship between hikers' experience and their knowledge or practice of bear-safe behaviors.[12]

On the plus side, the graduate student, like the Wyoming Survey and Analysis Center, found that hikers' basic knowledge of bears was fairly high. More than 80 percent knew that both bear species lived throughout Grand Teton Park, were active both night and day, and could break into vehicles if windows were cracked open as little as an inch. Two-thirds of those surveyed did not feel vulnerable to a bear attack, though 31 percent reported having had or knowing someone who had a potentially dangerous encounter with a bear. A majority did believe that being within 100 yards (the distance required by greater Yellowstone parks) of a bear was risky but thought grizzlies were more likely to cause critical injuries than black bears. To protect themselves, only about a third of the hikers called out, clapped, or made noise to alert bears of their presence. Generally, hikers' perceived vulnerability tracks with the park's low record of bear-caused human injuries. People surveyed believed that most recommended bear safety behaviors were easy to follow, except for not hiking at dawn or dusk—40 percent found that to be a challenge, which could be due to trails being less crowded and/or hot during summer and because of competition with other things in their schedules.

Perhaps most interesting was that while 65 percent of those surveyed thought bear spray was an effective practice and 80 percent thought it would be easy to keep it handy while hiking, only 28 percent carried it. Other parks have noticed similar results, prompting concern that the high cost (typically $40 or more per can) or inconvenience of purchasing bear spray and the inability to transport it on airplanes may deter its actual use. Efforts to provide day or short-term rental options exist in Jackson, Wyoming, and elsewhere, but the locations are seldom convenient for many hikers. Summing up her study, Blotkamp recommended stronger warnings about hiking in the dawn and dusk hours, since two serious bear attacks on humans occurred in the early morning. She urged that informational messages be more direct, such as saying, "Bells are not enough in bear country" and "keep food within arm's reach."

The social science studies provide important baselines for the effectiveness of educational materials and efforts to keep people safe. Though the risk of bear-human encounters is still quite low, the outcome can be severe, even fatal, for both people and bears. As of 2023 the park, along with local partners, reportedly planned to again evaluate bear information materials and signs as they embrace new designs and flexible messages to promote a bear-smart Jackson Hole.

Some people view additional research as an ever-present delaying tactic, or a means to spend more funds supporting scientists' pet projects. While that may sometimes occur, I feel compelled to defend the need for good research that does *not*, in most instances, determine all the facts about how the world works, or find definitive answers to anyone's burning questions. Science *is* about asking questions, exploring (sometimes experimenting with) alternative possibilities, and concluding with an explained level of certainty what, at the outlined time and conditions, the investigators discovered. As things change—the physical, biological, and/or social environment or the

FIG. 15. Grizzly bear near trap, remote camera photo. Courtesy U.S. Geological Survey Interagency Grizzly Bear Study Team.

methodologies available to seek answers—so might the discoveries. For instance, older literature described bears, even grizzlies, as mostly plant eaters well into the 1990s. Then the story changed. Researchers learned that in picking through bear scats they were missing the meager evidence of fully digested meat until other methods augmented their understanding of dietary dishes. Numbers of elk and bison and cutthroat trout (for a time in Yellowstone National Park) increased, making them more available to be preyed or scavenged upon. Bear numbers expanded and the animals learned or relearned how to catch such food, sometimes in full view of visitors lucky enough to see grizzlies chase elk calves in the Willow Flats of Jackson Hole and elsewhere. DNA sampling has allowed researchers to investigate relationships between individual animals that could only be presumed from observations in the early days of bear research. And it has led to methods that "capture" some types of information without needing to catch animals.

It has long been true that some people question the need for researchers to capture animals, or the methods they use to do so. I understand and respect such concerns, having taken part in catching and immobilizing bears and other wildlife. As with studies of food habits, research has evolved from the days of the first simple tools used to track grizzly bear movements. Tracking collars, for many years now, have been expandable to allow for animals' seasonal growth and built to fall off if too tight or are programmed to come off after a defined duration. They are no longer limited to VHF radio frequencies but often use satellite positioning systems to follow target animals. Traps have been improved to reduce the animals' risk of injuring themselves. Wildlife veterinarians train staff who must capture bears these days, and trappers share their best practices. Handlers choose newer immobilizing agents to cause less stress and use modern instruments to monitor animals' vital signs. There is scant record of research captures resulting in visible harm to bears, none in Grand Teton National Park or the parkway. As techniques evolve and become available to field biologists, the trend toward using non-invasive methods, such as remotely operated cameras, to "capture" animals have grown with researchers' support.

Whatever their tools, bear researchers certainly are hands-on outdoor practitioners, not desk-bound computer jockeys (at least until required to enter and analyze data, typically while bears are denning). While wildlife management specialists, in and out of parks, are often highly competent at animal capture and immobilization, research trappers employed full-time to catch bears for study are, in my experience, particularly memorable characters. They are certainly not in it for the pay. To spend nearly every day for seven months a year trying to trap an ursid that could kill you with jaws or claws, but that mostly just tries to avoid capture, attracts a special kind of worker. They drag smelly, mangled road-killed elk, deer, moose, and the occasional whatever else off the highways to use for bear bait. Some concoct their own scent lure to add as a bear attractant, needed even for wire "hair snares," cooking up incredibly putrid mixes on a stove at home in a pot never again fit for human cooking. Long-timers in an ecosystem come to know the traits of certain bears and pride themselves on the tricks they learn to try to outwit trap-savvy animals. They spend days or weeks on end out in the woods, traveling by foot or by boat or, more likely, using experienced horses and mules, which are remarkably helpful in carrying gear and bait while providing an added measure of safety if one needs to make a quick retreat in dicey situations. I have known trappers to camp not just in tents but in trees and even in bear traps adjacent to those intended for bear, as an extra but cold and uncomfortable measure of safety. The disincentive to turn too many clothing items into "field gear" and the time and distance from a shower only contribute to the auditory aura of trappers. More than one grizzly bear trapper who had gone to college or previously worked with my biologist husband attended our wedding. I remember thinking that one cleaned up surprisingly well and that another had obviously come straight in from the field that day—we could distinctively smell him in the receiving line after the ceremony.

People may think that getting paid to study and trap bears and other charismatic wildlife is prestigious or fun. Well, the work is certainly not mundane. It is an unpredictable venture, fraught with danger for both humans and bears, and every experienced trapper I've known longs, more so over time, to *not* have to handle the animals. But when they do, they show remarkable care, even devotion, unseen by the public and unheralded by the press, as they know better than anyone the awesome responsibility they hold to protect their study subjects. Even the snaring of hair to capture bear DNA for genetic, disease, and population monitoring involves packing gear and smelly bait for miles over rough terrain. Field research involves long, tedious hours of repetition if not boredom, in unpredictable weather and sometimes downright dangerous conditions, punctuated by moments of excitement and occasional new discovery.

And the time and attention to detail paid by investigators and statisticians who compile and analyze amassed data, then write and present it to their peers and to interested citizens, also merit respect—rare is the researcher who prefers that less appealing but crucial step in the scientific process. I admire and support those—wild animal veterinarians and trappers, biologists who identify and map plants from the valley floors to the mountaintops and who pick through animal scats to meticulously identify what was consumed, sociologists and scholars of wildlife-human dimensions, cone and bone and DNA collectors, laboratory technicians, authors of publications, and more— who devote their professional lives to improving our knowledge of bears and other aspects of nature, so that we and they can marvel at what has been and is yet to be discovered in the Tetons and elsewhere in wild ecosystems.

Bears in the 'Hood

By 2012 grizzly bear #399 was increasing her movements into the southern part of Grand Teton National Park, especially the Moose-Wilson Road corridor, in late summer and fall when chokecherries and hawthorns were abundantly available. Interested citizens weighed in on whether the Park Service should "improve" (by widening and/or paving) the increasingly well-traveled route in the park's southwestern corner, and how they should manage bear-human conflicts along the narrow road lined with vegetation. The popular and prolific mother bear moved through the area at various times of year without my seeing her, despite my home and office being at the north end of the corridor. But I awakened one cool Sunday morning in May 2014, at my government residence in the Moose housing area, to find an adult grizzly paw print on the window of the mud room door, and tracks of mother and cub in the frosty driveway. Other than a rather atypical indoor bark from our dog the night before, my family had neither seen nor heard anything of bears in the neighborhood. When I reported the track to the bear management specialist, she informed me that no one else had reported seeing bears or their sign in the housing area that night. We joked about #399 singling me out to leave a message about who was really in charge of "bear management."

Comparing historical accounts to present-day Grand Teton Park often reminds me how, in ways, little has changed. In the mid-to-late 1970s, park plow driver Barry Alexander lived with his family in the housing area at Colter Bay, where they would sit on the porch and watch bears. He recalled a Lodge Company cook who brought home suet and fat and poured it over stumps behind the company dorm; by 2:00 p.m., four or five bears would be in the housing area with kids playing nearby. Now and then, if he left his garage door open, he'd find bears inside, always black bears that nobody worried about.[1]

People should expect bears and other wildlife to wander at will in parks, with little regard for visitor or employee facilities. Of course, managers today would not tolerate blatantly placing bear attractants outside, especially when done by employees. But history before laws such as the National Environmental Policy Act has saddled parks with developments built for visitors very near natural attractions—riverbanks, lakeshores, canyon rims, sequoia groves, geyser basins, and grand scenic views. Agencies also built administrative areas nearby to be convenient for workers, maintenance, and emergency response. There was little recognition or conscious avoidance of sensitive habitats and species. Decades later, a robust body of research has demonstrated that roads, road density, and developed areas are the most important factors affecting grizzly bear survival, as they provide more chances for disturbance and encounters that could result in the death of bears.[2] Bear use, at least when habitats were not saturated, was not expected near developed areas and vehicle routes, even in sparsely roaded Yellowstone.[3] This led the ecosystem's managers to establish standards to maintain what is termed secure habitat.

Bear avoidance of human-use areas is not, of course, an absolute. It's too much to think that bears will ignore the temptation of huckleberry patches near String Lake or at Signal Mountain Campground, even at peak summer occupancy, setting the park up for a continuing need to educate people and manage human activities. The present-day park headquarters at Moose, being mostly in sagebrush and grass, gets fewer sightings than other developed areas, although observers sometimes see bears moving through the trees near the Moose Entrance Station or along the Snake River. The original headquarters near Beaver Creek, where park employee housing remains today, is in the forest edge at the base of the Teton Range, a travel corridor for bears, elk, and other wildlife. Similarly, animals traveling along the shores of Jackson and Jenny Lakes can move right through the lodge complexes and campgrounds sited near the waters. Since developments in the park and parkway exist in good habitat, we should expect wildlife to be in the neighborhood.

However, bears that linger in areas regularly occupied or visited by humans have long made bear managers uncomfortable. Under the 1986 *Interagency Grizzly Bear Guidelines*, developments occupied by residents or visitors in lodges and campgrounds fall in Management Situation 3: areas primarily managed for people, where bears will be actively discouraged from remaining.[4] Bears may wander through, but rangers trained in hazing techniques, such as yelling or using other noisemakers and firing cracker rounds, try to discourage them from staying. Within the Grizzly Bear Recovery Zone, the managers' prescription for no net habitat loss restricts agencies' ability to build new roads or developments. Also, the *Master Plan* for Grand Teton National Park called for no new developed areas; thus, while there's some flexibility for infill, there should be no new Flagg Ranch or Colter Bay–type visitor use areas.[5]

Scientists and managers aiming to prevent conflicts have come to distinguish between bears that are "habituated" and those that are "food conditioned."[6] (Think of your pet expecting to be fed in the same dish at the same time each day.) Bears learn to find easy chow during mealtimes at campgrounds, anywhere they smell meat on the barbecue, and at fish-cleaning stations, just as they once did at the dumps. Bears presented with such temptations can learn aggressive approaches such as chasing people from their picnic tables. While this may call to mind Yogi Bear cartoons, it's serious when a real bear is in *your* yard or campsite.

Through clever and persistent investigation of locks and latches or with brute force, bears can break into beehives, storage sheds, kitchens, and vehicles—thus the widespread efforts to prevent them from obtaining our food and garbage. Bear-caused property damages, close encounters, and of course human injuries are always of concern. A bear manager's ever-present fear is that an animal will cross an invisible but real line between the merely nerve-racking and the intolerable.

FIG. 16. Black bear looking into window at Murie Ranch residence. Photo by Christen Girard. National Park Service photo.

Decades ago, park superintendents and national forest supervisors enacted food storage rules for lands they managed; it has been more challenging in local counties and towns. By the mid-1980s the Wyoming Game and Fish Department commonly recommended that, prior to being approved, residential projects like the Brown Subdivision on the park boundary near Pacific Creek should require bear-proof garbage containers.[7] In the 1990s Teton County commissioner Sandy Shuptrine so consistently called for that proactive requirement when considering new developments that her peers informally called it the "Shuptrine amendment."[8] Multiple agencies' biologists urged the county to require bear-resistant food storage outside the park, but it took until 2008 for the land development regulations, or LDRs, to do so, and the rule applied only from April 1 through November 30 in an area termed "Conflict Zone 1," based on the state's bear location data. Residents had until 2010 to implement the new rules. More vexing questions remained about whether to discourage or restrict food gardens and fruit-bearing plantings, bird feeders, even trash composting in bear-occupied areas. Lively debates continue as entrepreneurs and conservationists seek creative solutions to help people live in bear country and prevent animals from becoming food conditioned.

Habituation, on the other hand, pertains to bears' ability to adapt to human proximity and go about their business without receiving food rewards. Such behavior can vary depending on an individual animal's tolerance and on the situation.[9] If bears ignore, or appear to ignore, people, this doesn't ensure that they are unaffected by the human presence. Research in Sweden found that brown bears may tolerate people and their developments, especially during berry season, and yet have elevated heart rates, an indication of stress.[10] Nevertheless, the trend toward distinguishing between food-conditioned and habituated animals led to some changes. Until the 1990s Yellowstone Park worked to haze all bears from roadsides or move them to backcountry areas. It seldom deterred the bears for long. Thus, rangers began tolerating roadside non-food-conditioned bears and using staff to help people stop and safely watch. Grand Teton adopted Yellowstone's more tolerant practice when its first habituated grizzly turned up feeding along a road in 2004.[11] This practice has resulted in more and more bear jams in the parks, such as along the Moose-Wilson Road—when berries are ripe, reward outweighs risk for many bears.

By 2005 Grand Teton Park was engaged in transportation planning, which included the historic seasonal road connecting the park with Teton Village—an expanse of lodges, shops, and restaurants at Jackson Hole Mountain Resort—and the community of Wilson, at the base of Teton Pass. There was interest in paving a rutted and bumpy 1.5-mile section, and in potentially even opening the road year round. Pathway advocates urged construction of a route for bicycles and other non-motorized uses, separated from the road by up to 150 feet.[12] Prior to transferring to the park the last of his JY Ranch, midway between Teton Village and Moose, associates of the late Laurance Rockefeller had built a new welcome center and revamped trails to Phelps Lake, in keeping with their vision of encouraging low-density natural experiences. Park managers recognized the intense growth in road use while pointing out the high scenic and wildlife habitat value of the corridor, which also holds significant archaeological resources. Resource specialists and others were concerned about the effects of new construction, essentially widening the corridor in which bears and other animals would be displaced by user traffic. What would normally be a local issue was elevated to the National Park Service director and then-U.S. senator Craig Thomas by members of a local group who referenced Rockefeller's concerns over development pressure in the area. With typical passion across a spectrum, valley residents spoke out about various issues: needing more roads or mass transit systems to reduce congestion and commuting times; the health and reduced emission benefits to users of bike paths; and the value of protecting mature forests, prehistoric sites, and wildlife habitat near the Snake River. After considering nearly 2,700 comments on the draft parkwide transportation plan, in March 2007 the Service approved a version that would allow some realignment of the Moose-Wilson Road and, as part of an estimated 45-million-dollar five-year project, construction of three miles of pathway, separated by 150 yards, in the woods.[13]

At the time, the corridor seemed inhabited by only black bears. Rangers assigned to what became the Laurance S. Rockefeller Preserve proposed a seasonal trail reroute from the contact station for visitors to walk more safely to Phelps Lake when bears forage in the thick brush along Lake Creek. I have seen three-hundred-pound animals sit on the asphalt surface of the Moose-Wilson Road and pull fruit-laden hawthorn branches down from the steep hillside just off the pavement, then pluck berry clumps like grapes. In the early years of the Wildlife Brigade, a volunteer directed visitors passing by, including a bride in a long white wedding gown, watching black bears gorge themselves.

In late April 2008, while the park energized its bear management program, grizzly bear #399—yet to have her own social media sites—roamed southward from her usual haunts near Pilgrim Creek and Willow Flats. Over the course of a spring weekend, I watched the mama bear and her three two-year-old cubs grazing and eating a carcass high on Blacktail Butte east of Moose, not anticipating how far and how often she might push the bounds of her home range in years to come. By May of the following year, state biologists were aware of two grizzlies in the southern part of the Tetons, one in Teton Canyon on the west side of the mountains and one south of the park near Teton Village.[14] But the fledgling Wildlife Brigade still focused their time on the northern half of Grand Teton National Park, where most wildlife jams occurred.

Then grizzly #399, her daughter #610, and their total of five cubs turned up along the Moose-Wilson Road in October 2011, and the younger mother charged a ranger patrol car. More staff were called to respond, but their inability to safely keep watchers anywhere near the required 100-yard distance from bears, or to safely direct traffic on foot as they typically do, caused the superintendent to prohibit vehicle traffic in the corridor. By 2012 the park had adopted the practice of closing the road when grizzly bears were present. At first it allowed visitors to walk in, then it decided to limit that, too, due to the risk of bear-human encounters. Despite an evolving array of practices for when and how to implement closures, there and elsewhere, shutting down a regularly used roadway was an interim solution at best. Park staff, with no shortage of well-intentioned outside input, considered an array of management measures—hazing animals, removing vegetation that provided an attractive nuisance, road redesign, escorting or phasing entry for vehicles through the corridor, and a mind-numbing variety of parking-deterrence tools like traffic cones, boulders, and bollards. Little seemed fiscally feasible, socially acceptable, or compatible with the Park Service mission to conserve resources while providing for public enjoyment.

For a decade, Grand Teton staff first considered how to implement the approved transportation plan then, due to changing conditions and high public interest, undertook a new comprehensive Moose-Wilson Corridor plan and environmental analysis. Throughout, providing for wildlife and safe viewing opportunities was a major aim, but in just five years, grizzlies had joined black bears in the mix of resources to consider. Overlapping the planning period was a new social science project to describe wildlife jams. They varied from 5 to 120 minutes in length, concentrated in two areas of the road corridor. Researchers observed that visitors on average got within fifty meters or yards of bears, about half the allowed distance, but some were as close as five meters (fifteen feet). Fortunately, in only 17 percent of observations did bears respond with "alarm behaviors" such as looking up or walking/running away, and no one was injured. Although scientists did not suggest that the volunteers were ineffective at educating visitors, the presence of Wildlife Brigade workers did not affect the duration of jams or even the distance of bear-human separation.[15] A new plan, approved in 2016, called for road paving and minor realignment, no separate bike path, reduced speed limits, and a maximum of two hundred vehicles in the corridor at a time, to be achieved by some form of phased entry. Managers could remove fruit-bearing shrubs if they planted more elsewhere to prevent a net loss of such habitat.[16] Implementation of that plan began in 2022, and eventually the road redesign could reduce the proximity and risk of bear-human encounters. Nevertheless, expectations are for continued intensive management of wildlife jams along the Moose-Wilson Road.

Grand Teton National Park has become a place renowned for offering a good chance to spot a grizzly along a park road. Family groups and subadults have dominated grizzly bear jams, which could be partly because they perceive the threat from humans to be less than the danger from adult male bears.[17] It surely ups the "wow" factor for watchers, as well as the potential danger to visitors and managers, when multiple animals are present, especially protective mothers and their young. And it has prompted ongoing discussions about whether to allow bears to become habituated to humans.

Park biologists have attended national workshops and, with their colleagues in the ecosystem, coauthored papers on the pros and cons of habituation. Benefits are that bears have access to roadside habitat from which human activity otherwise might displace them. In Yellowstone Park alone, that amounts to acreage the size of two adult female grizzly bear home ranges—about fifty square miles (36,000 acres) when the cubs are smallest and move the least quickly, up to 150 square miles (96,000 acres) as cubs age to subadulthood.[18] There is some evidence that habituated brown bears in Alaska, Montana, and greater Yellowstone are less likely to threaten or injure hikers and bear watchers.[19] Roadside bear viewing is certainly popular with visitors and photographers and has been an economic bonus to park concessioners, tour operators, and other businesses in gateway communities.[20] Between 2008 and 2018, Grand Teton staff and volunteers managed at least 1,585 black bear and 1,369 grizzly bear jams. These jams provide an opportunity, even a need, to capture visitors' attention and educate them on bear ecology, behavior, and conservation.[21] More than half of people surveyed in Yellowstone National Park valued bears more because of roadside viewing, and more than 60 percent said it inspired them to accept habitat protection and some limits on both development and recreation to protect bears.[22] This is likely also true in Jackson Hole.

Canadian bear researcher Stephen Herrero and others have neatly summarized the negatives of bear habituation.[23] There are more traffic jams and associated inconvenience; some travelers dislike being delayed. People who see habituated bears may assume that all bears will tolerate their proximity without causing stress to the animals or danger to watchers. In Canada habituated bears are more likely to be killed by poachers, and this could also be true in the United States. Although a 2018 analysis did not find an increased rate of road-killed bears since Grand Teton and Yellowstone National Parks began managing bear jams, habituated animals may be at higher risk of being hit by cars or being killed outside the parks. When grizzly #615, from one of #399's first litters, was killed outside Grand Teton Park by a hunter fearing for his safety, the bear's tolerance of human proximity was thought to be a contributing factor, and several other cubs in #399's lineage have been killed in management actions. Would discouraging habituation have kept these bears from dying? Or would it have prevented them from surviving? No one knows. But habituated bears may be more likely to eventually receive human handouts. If they expect or demand even more food, this could cause property damage or injury to humans, leading the public or decision-makers to call for an animal's removal—thus the recurring watchfulness of bear managers.

Managing habituated animals takes money and time, lots of time. As does preventing, as much as is possible, human-bear conflicts in neighborhoods, in high-use visitor areas, and even in the wild backcountry. Grizzlies now fully occupy Grand Teton National Park. The presence of roadside bears there and elsewhere in greater Yellowstone has boomed and shows no signs of slowing down. Sustaining a long-term commitment to managing people and habituated bears is described as the parks' biggest wildlife management challenge. Fortunately, individuals and organizations have stepped up in tangible ways to help meet the challenges of living with bears in Jackson Hole.

17

Making a Difference

One summer morning, while staying at Colter Bay Campground, my hus-
band and I made our way to site J197, where a nice couple was nearly packed
up and ready to check out. They did not object when I asked if we might look
at the food storage box anchored in the ground beside the large pull-in camp-
ing spot. I explained that some of the boxes were donated, through the Grand
Teton National Park Foundation, in honor or memory of someone, and that the
one in this site was supposed to have my name on it—a generous gift from the
foundation upon my retirement from the park in 2019. I knelt to read the small
metal plaque installed on the brown steel box and, even knowing what to expect,
having seen and occasionally helped install commemorative plaques on other
boxes in my working years, I was surprisingly moved by the tangible token. It
took me back to when we started the program, in 2008, with the goal of placing
a box in each of the developed campsites and other hot spots in the park. And it
reminds me still of how, in the passage of not much time, so much progress has
been made by the combined efforts of numerous people and organizations who
have contributed to the conservation of bears in Grand Teton National Park
and the Rockefeller Parkway.

There are many ways for people to influence bear conservation. Jackson Hole
is renowned, at least since the onset of the modern environmental movement

FIG. 17. Bear box at Colter Bay Campground. Photo by author.

THIS BEAR BOX WAS FUNDED THROUGH
GRAND TETON NATIONAL PARK FOUNDATION

IN RECOGNITION OF
SUE CONSOLO-MURPHY

GRAND TETON NATIONAL PARK
CHIEF OF SCIENCE AND RESOURCE
MANAGEMENT
2003 - 2019

FIG. 18. Plaque dedicating bear box. Photo by author.

in the 1970s, as an area that values wildlife for its intrinsic value as well as for economic benefits related to the popularity of wildlife photography and viewing. Local and national organizations have for decades been engaged in or supported conservation causes. The National Parks and Conservation Association, the Jackson Hole Conservation Alliance, the Greater Yellowstone Coalition, and others have actively promoted habitat preservation, environmental planning, and other initiatives to benefit bears and other wildlife. Countless people produce or contribute to internet blogs and social media sites sharing well-intended and mostly accurate bear information. Grizzly bear #399 has "her own" Facebook pages with huge followings. While it's hard to assess the contribution of online fandom to wildlife conservation, it certainly reflects public interest in individual animals and, I hope, in maintaining wild populations and their habitat. Local and even national media have featured hundreds of bear stories from the earliest settlement of Jackson Hole to present day. Not infrequently, they have offered opinions on bear management issues, which has no doubt influenced citizens and elected officials to act.

While recognizing that this is just a sample of good efforts and the people behind them, I highlight several programs that have been especially focused in and around Grand Teton National Park. Bear Wise Jackson Hole is a partnership involving the Bridger-Teton National Forest, the park, the Wyoming Game and Fish Department, and the Jackson Hole Wildlife Foundation (a local nonprofit organization of which I was a nonvoting board member representing the park for six years). One of the Wildlife Foundation's four focus initiatives, Bear Wise promotes resident and visitor education as well as use of bear-resistant trash containers. Even before Teton County, Wyoming, enacted its first requirement for bear-resistant food and trash containers in 2008, the group ordered Interagency Grizzly Bear Committee–approved trash cans and offered them for a reduced price to locals. By late 2023, Bear Wise partners estimated compliance at 98 percent in areas of the county where regulations had long been in place, mostly north of Jackson, and 70–80 percent where the rules had more recently been enacted in the town and county. The foundation has a long-standing campaign to "Give Wildlife a Brake," sponsoring roadside speed-readers and other messages promoting slower and attentive driving to deter vehicle-wildlife collisions. In 2019 the community passed a special excise tax measure to help fund wildlife crossings on busy roads.[1] The Wyoming Game and Fish Department and local businesses have sponsored bear spray give-aways, providing the $40+ canisters for free to some lucky recreationists each year. And Bear Wise partners collaborated to purchase an educational trailer, often stationed at a popular overlook near Jackson Lake Lodge in Grand Teton National Park. In 2022, for example, park staff or volunteers dedicated mul-tiple hours on forty-five days talking to visitors, sharing bear safety and ecol-ogy information illustrated with display animals, skulls, hides, and monitoring equipment stored in the trailer. The popular mobile tool is sometimes moved to town or other valley locales for special events, such as Earth Day celebrations.

Another major partners' effort helped the park reach its goal of hav-ing a bear-resistant food storage box in every developed campsite as well as selected high-use picnic spots. Grand Teton National Park concessioners—the Grand Teton Lodge and Signal Mountain Lodge companies, which have been key cooperators in ongoing efforts to educate visitors and enforce food storage rules—provided about $750,000 over fifteen years through franchise fees, a portion of their profits returned to the government to use for park-approved improvements that benefit visitors.[2] The Grand Teton National Park Foundation is the park's major nonprofit supporter and, since its incep-tion in 1997, has helped the park build the Craig Thomas Discovery and Vis-itor Center, acquire 640 acres of land on Antelope Flats and the Moulton Ranch Cabins on Mormon Row, improve trails and educational displays at Jenny Lake, and redesign Snake River boat launches and overlooks. The foun-dation created a popular campaign for its members and donors to "adopt a bear box," to which a commemorative plaque could be attached, as illustrated in the opening to this chapter. Donors have bought boxes in honor of loved ones, and some have purchased multiple boxes in celebration of family or friends' birthdays. When the effort began, the steel boxes could be acquired and installed for less than $1000 each, but in the 2020s the cost to purchase and install each box more than doubled due to inflation and limited sup-ply chains during the COVID-19 pandemic. Nonetheless, foundation-raised funds paid for purchase and installation of 694 boxes, more than half of the 1,069 boxes added by the end of 2023. Every front-country campsite in the park had a bear box. With the foundation's continued support, staff plan to replace fifty-four older boxes in the Jenny Lake Campground and add more boxes at popular lakeshore areas.[3]

The Grand Teton National Park Foundation has also supplied crucial funding for bear monitoring and management, including almost $233,000 for research on how the elk reduction program might affect grizzlies and more funds for bear-tracking collars and "Be Bear Aware" educational efforts. Bear jams, more than anything else, prompted creation of the park's Wildlife Brigade, and from 2012 to 2020 the foundation provided $350,824 to help support them. By 2023 thirty-one Wildlife Brigade volunteers assisted a permanent park supervisor and two seasonal rangers. In its first fifteen years of existence, dozens of Wildlife Brigade returnees and new recruits donated their time—more than 12,000 hours in 2022 alone—working outdoors in all types of weather conditions, risking their own safety with bears and vehicles so near, and occasionally being treated with less respect than they deserve. It's worth it, according to Jeff Willemain, who began volunteering with the Wildlife Brigade in 2015 and has served as chair of the Teton Park Foundation's board. He fondly recalled a family of grandparents and grandchildren who had wanted to see bears all their lives. When he guided them in safely watching a black bear and cubs on the Moose-Wilson Road, the elderly woman cried with joy. Other visitors have hugged him, they were so happy to see a bear.[4] Jeff described the energy in the foundation boardroom when members discussed support for the bear box and brigade as being inordinately bigger than the amount of money provided—it reflected members' passion for bears and for the brigade being ambassadors for wildlife in the park, helping visitors safely have the experience of their lives.

Bear boxes and wildlife jam attendants that barely existed in Grand Teton Park two decades ago are now part of the scene, which blends into the grand views. As does one of the most notable changes on the landscape, likely unnoticed by most visitors. It occurred in the early part of the twenty-first century, prompted by the string of bear-caused cattle killings in the mid-1990s. Wyoming Game and Fish Department habitat biologist Steve Kilpatrick, now retired, recalled working with a sheep grazing permittee in the Grey's River area south of Jackson Hole in about 1997. The rancher offered to close his allotment and move his sheep for about $28,000. That, Steve said, "got him thinking," and by 1999 or 2000, while talking with the manager of the Walton Ranch about grizzly bear–livestock conflicts on the Blackrock–Spread Creek allotment, he asked if there was any way the state or its partners could buy out the grazing rights.[5]

Apparently, the initial answer was, "for a million dollars!" But the ranch manager agreed to ask Betty Walton, the surviving ranch owner, who was struggling to figure out what to do with the long-term lease held by her and late husband Paul. Kilpatrick tried to raise money, an experience not in his portfolio. Fortuitously, Hank Fischer—who had served as a primary representative of the Defenders of Wildlife during plans to restore wolves to the Northern Rockies and who had moved to the National Wildlife Federation—called to pitch a then-fledgling program to offer ranchers economic compensation for retiring grazing permits on public lands in areas of heavy livestock-predator conflicts. Fischer and Kilpatrick teamed up, pooling the former's organizational and fundraising experience with the latter's knowledge of the local landscape and relationships with locals. Despite initial skepticism from district-level personnel, Bridger-Teton National Forest supervisor Kniffy Hamilton thought that she could justify closure of the allotment, given Forest Service management responsibilities for threatened and endangered species. The National Wildlife Federation developed an agreement that identified the financial compensation and associated fundraising needed; it also required the permittee to waive the grazing permit back to the forest. The Forest Service could then put the area into one of three categories: 1) a vacant allotment, the least secure option in the long run; 2) forest reserve, which would be used only in case of catastrophic events; or 3) closure, a permanent change that requires more long-term planning and compliance with the National Environmental Policy Act.

Kilpatrick worked with a Walton Ranch attorney to negotiate the agreement. In about two years, Fischer and others raised funds—the agreement kept the amount confidential, but the National Wildlife Federation reported that they presented a check to Betty Walton in August of 2003 for $250,000—to vacate the Blackrock–Spread Creek allotment.[6] It was only the second grazing permit buyout in the federation's program and the largest such arrangement, securing 87,733 acres from livestock grazing in Management Situation 1 and 2 grizzly bear habitat.[7] Follow-up efforts later removed cattle from 178,000 acres of the Bacon–Fish Creek allotments immediately south, extending into the Gros Ventre Range. For that buyout, Kilpatrick remembers that some of the same donors helped and also created a "buy an acre" campaign to encourage smaller financial gifts. Eventually more than two hundred supporters contributed.

Steve Kilpatrick described himself as the "boots on the ground" for these efforts, and he credited Hank Fischer for partnering with him to handle the politics and fundraising.

> It was the most rewarding and successful thing I got to do in my career, so far above anything else I ever did for wildlife . . . Grand Teton National Park isn't very big, but this change in management of the two allotments almost doubled the effectiveness of a small national park in terms of wildlife habitat. Grand Teton has 310,000 acres and the combined allotments were more than 265,000 acres. How different it would look today—the number of bears and livestock killed—if we hadn't done it?

After the grazing buyback was completed in the Spread Creek area, Steve worked with the Jackson Hole Wildlife Foundation on another of their key initiatives—removal or modification of rangeland fencing to make it more wildlife friendly. Often this involves replacing the bottom strand of barbed-wire fencing with smooth wire and ensuring that it is at least 16 inches above the ground to permit pronghorn and deer to easily cross under. Top wires should ideally be no more than 42 inches high and separated from the strand below by 10 inches to prevent elk and deer from entangling their legs while jumping fences. In the case of the former Walton Ranch allotment, fencing was no longer needed. So Steve coordinated volunteers from the Jackson

Hole Wildlife Foundation and the Lander Chapter of the Rocky Mountain Elk Foundation to remove fences from both park and forest land in the Elk Ranch Reservoir area and eastward to the Continental Divide. Jackson residents Chuck and Carol Schneebeck, who after moving to the valley in 2000 led fence-pulls for the Wildlife Foundation for six years, waited until a wolf pack had vacated their den in the former park pasture each August. Their volunteer crews worked upstream from the Moosehead Ranch along Spread Creek, pulling fence by hand from gnarly willows, and through the sagebrush flats directly south of Wolff Ridge, removing a spur that led to the dam that then still existed on the stream. Then they continued east and north, pulling fence beyond the Uhl Hill–Elk Ranch reservoir and ending at the edge of the Buffalo Fork River. They got permission to use all-terrain vehicles (ATVs) and obtained a wire-winder to pull downed fence into huge rolls to be hauled out later. The volunteers removed nearly eight miles of derelict fencing from Grand Teton National Park during that project, along with another eleven and a half miles between Gros Ventre Junction and the town of Kelly, where cattle brought from private ranches south of town typically grazed early in the season before moving to the Elk Ranch. During Wildlife Foundation–sponsored fence removal or modification events in the summers of 2004–2006, Schneebeck estimated that ten additional miles of barrier came down on the Bridger-Teton National Forest.[8] The end result, while poignantly marking a trend away from cattle ranching in the heart of Jackson Hole, made a tangible difference. As Schneebeck put it, "the best result—nothing was left" except thousands of unfenced, ungrazed acres to benefit bears, native ungulates, and other fish and wildlife, and the scenic landscape of the Tetons.[9]

The history of bears, and bear management, in the Tetons has evolved over the course of the park's nearly hundred years from a place where bears were hardly an attraction to one that regularly makes the list of top sites for wildlife watching, a "grizzly park." The success of programs to help people live and work and recreate safely in bear country, and to conserve wild populations of black and grizzly bears, will continue to rely on many hands and minds to keep it so.

18

Celebrity

It was not your average night in Jackson, Wyoming, as the oft-touted "world's most famous grizzly bears" came to town. After a late-season foray out of Grand Teton National Park, where she was most often seen, grizzly bear #399 and her atypically large litter of four cubs—then half her size and forming a veritable herd of bears—made their way back north from the Snake River Canyon and other locations south of Jackson where, unfortunately, they'd gotten into unsecured beehives, compost, livestock grain, and garbage.[1] More than concerned, wildlife managers ordered traps set for the bears in hopes of capturing and radio collaring at least some of the family so they could track the animals' movements and keep them out of harm's way.

On November 6, 2021, biologists from the U.S. Fish and Wildlife Service, the Wyoming Game and Fish Department, and the Interagency Grizzly Bear Study Team successfully collared two young male bears, helping the watchers monitor and "steer" the animals, who chose to move right through town for a few days ahead of the coming winter. Jackson Police Department officers had been around too, directing traffic and, when necessary, cautioning town residents to keep their distance. Though swarming with visitors in the height of both summer and winter, in the off-season Jackson is relatively quiet, like any other town of just 10,000 permanent residents. It was below freezing, and a cold front was bringing snow to the valley at about 10:00 p.m. on Tuesday, November 9. Security cameras mounted outside the police department captured little automobile

traffic. But one did catch five bruins walking right through the town hall park-
ing lot. Police Chief Michelle Weber chose to post the video footage, aptly backed
up by audio of the old Yogi Bear cartoon theme song ("smarter than the average
bear"), on social media to make sure locals knew to be alert and put away food
and garbage so as not to attract the animals.

One morning during their fall foray, the chief, who lived near the southern
end of town, saw five sets of bear tracks in the snow that had fallen in her yard.
When I chatted with her in July 2022, Chief Weber said that, despite often ven-
turing into the park on her time off to boat on Jackson Lake, she had not previ-
ously chanced to see the famous family of five, so it was pretty cool that #399 had
come to both her house and her office.[2]

Front-page news in the *Jackson's Hole Courier* one August day of 1930 told
of an animal that "parked his royal grizzly carcass in the middle of the road"
over Teton Pass, stopping cars and entertaining tourists in midday.[3] It was a
rare sighting then and for most of the following seventy-five years, until grizzly bear #399 appeared on the Jackson Hole scene. I struggled with what to
write about the renowned ursid, fully aware that some readers might assume
she is central to any story of bears in the Tetons. As an eyewitness, sometimes
a player in deciding her fate and that of her progeny, and certainly a responsible party in helping shape Grand Teton National Park's response to her, I
cannot view her with objectivity any more than others can. Her personal history has been told, in greater detail than I choose to repeat, in books by Jackson resident Tom Mangelsen, who provided eye-popping photographs, and
Todd Wilkinson, who wrote the text. Searching back issues of local newspapers provides more than a snapshot of the Tetons' most well-known bear
from 2006 forward, along with an interesting array of viewpoints in letters to
the editor, opinion columns, even paid ads bemoaning "TOO MANY GRIZ-
ZLIES."[4] And there is no end of information, fact-checked or not, available
on the internet.

For readers who might not know or who want a refresher, Jackson Hole's
most famous grizzly was born in 1996—who knows where—to an unidenti-

fied mother. She was five years old when first caught, in late August of 2001, by biologists from Grand Teton National Park and the Grizzly Bear Study Team, from whom she received her sequential radio-collar number. Of thirty-six individual bears trapped in research efforts (rather than for management reasons) that season, only two females were caught in the park, she near Arizona Creek north of Colter Bay, and twelve-year-old bear #179 farther south near Spread Creek. DNA analysis revealed no relationship between the two. Bear #399 was not averse to entering traps early in her life and was caught again seven times in or near the northeastern part of the park by the end of 2005. When handled in July 2004, at age eight, she evidenced having borne and lost a cub, but she had none with her when captured the following year. She would not enter a trap again for a dozen years. She still carried a working radio collar and at least one colored ear tag when she emerged from her winter den in spring 2006 with three cubs-of-the-year, but she cast the collar before she denned the following winter.

Her frequent appearance along Teton Park roadways caused her to become widely seen, photographed, and sought out by wildlife watchers. Her involvement in a seldom-documented case of cub adoption only added to the mother bear's renown. In 2011, #399, identified by her facial scars and a red ear tag, emerged from her den with three cubs-of-the-year, while her daughter #610, age five and bearing a yellow ear tag, had two cubs in her first litter. Around July 21, observers noticed three cubs accompanying the younger mother, while #399 escorted only two. In spite of an initial false accusation that biologists had caught the adult bears and changed their ear tags or mixed up the cubs, this was a natural occurrence. It was one of several known instances, like a previous event in Yellowstone National Park in which a daughter had taken on her mother's offspring.[5] Such exchanges could be the result of male bears or wolves harassing bear groups, or of the cubs intermingling when their mothers interact, but usually the reason for such adoptions is a mystery to us.

Between 2006 and 2021, grizzly #399 had seven litters and a total of seventeen cubs, eight of which were gone by 2022: two were killed by vehicles, one was killed by a hunter outside the park who surprised the bear and

FIG. 19. Grizzly bear #399 and cubs in snow. Photo by Gary M. Pollock.

defended himself, and five were removed for management conflicts, all but one outside the park.[6] Most notably, her 2020 brood of four cubs was twice the average litter size. At that time #399 was nearly but not the oldest bear in the Yellowstone ecosystem to have young; grizzlies #12 and #365 bore cubs at the age of twenty-five, and #125 and #541 matched bear #399 in doing so at age twenty-four.

The mother bear herself had surprised researchers, who had trapped and re-collared her once more in late summer of 2016. Within two years she had cast that collar, although she and her offspring left DNA in guard hairs snagged on wire hair snares and she remained recognizable to regular bear watchers, both government employees and private citizens, by her appearance and behavior. Frank van Manen, head of the Interagency Grizzly Bear Study Team, has called it rather remarkable that #399, even without carrying a radio collar for most of her reproductive life, was trackable so often.[7] For her first fifteen years, she centered her activity in Grand Teton Park, often being seen between the Colter Bay–Pilgrim Creek area and the area near the Snake River's Oxbow Bend or the Willow Flats adjacent to the Jackson Lake Lodge. She sometimes moved south to Blacktail Butte and the lower Gros Ventre River corridor, even to the Moose-Wilson Road or outside the park, especially during hunting season. But a 2021 sojourn around, through, and well beyond the town of Jackson seemed out of pattern even for her.

Seasoned researchers van Manen and Mark Haroldson of the study team have "seen" (using radiotelemetry data) other female grizzlies with cubs do what they call "big walkabouts" and offered that if more bears were followed throughout their long lives, biologists would probably see more of it.[8] Another bear that took a similar adventure was #533, a female that lived near Island Park, twenty miles beyond West Yellowstone, Montana. She moved one year with her yearling cubs and spent considerable time in the Electric Peak–Swan Lake Flats area of northwest Yellowstone National Park before returning to Idaho. The researchers suspected that #533 was related to well-known bear #264, who frequented park land between Mammoth Hot Springs and the Norris Geyser Basin, and speculated that the wanderer could have been returning to her natal range. They spoke of other older female grizzlies seen outside their traditional home ranges, and of why bears might take such a journey. In 2021, #399 was tending four yearling cubs, a rare enough occurrence. She was bigger than the average mother bear in the ecosystem, and since the nutritional demand for her and her growing family was substantial, it could have been to her advantage to expand the area in which they could find food and space. She could have been avoiding competition. Though large, at her age she might have lost vigor or dominance over other bears in her home range, yet she would still be compelled to teach her offspring how and where to find what they needed to survive.

The extended roaming beyond her regular haunts and the grizzlies' increasing presence ever closer to where people live was not universally popular, but, as expressed by one long-standing visitor-cum-local, "It's not the grizzlies' fault. They just shouldn't be living in the same environment as humans."[9] Bear #399 had, remarkably, kept out of trouble, with a low record of human conflicts throughout most of her life. The one human injury she caused was in defense of herself, her cubs, and their freshly killed prey, and park managers gave no consideration to moving or removing her in that instance. (While people injured by bears sometimes offer their opinions on the fate of animals that attacked them, decision-makers did not query the victim on what to do in response to that or any other bear attack during my tenure.) With the added stress of raising four strapping offspring and moving outside the more protected park lands, #399 and family repeatedly got into unsecured attractants. The U. S. Fish and Wildlife Service documented five instances of this in 2020 and seventeen in 2021. Worry for their and humans' safety culminated when the bears came through the town of Jackson. By sheer luck, the determination of many wildlife watchers, and some extra tolerance from bear managers—Grizzly Bear Recovery Coordinator Hilary Cooley, who worried that "the future's not so bright for these guys," admitted that they had been "more lenient" with the famous family—they made it through that episode to den again on the cusp of the new year.[10] As expected, in May of 2022 the cubs and their mother parted ways, and at least one of the litter did not fare well. Young male #1057 showed his habituated behavior too often and, after he failed to react when a resident tried to haze the animal off his front porch by firing warning shots, Wyoming wildlife managers trapped and euthanized the bear.[11] Fans wondered about the fates of the other subadults, and whether their mother—though having been seen in the company of a big boar during mating season—would ever emerge again to delight them, with or without young.

Grizzly #399 had long been big news in local and regional newspapers and online forums, from Facebook to locals' blogs, in which commercial and amateur photographers generously shared images and observations. Then, as if she weren't popular enough, "the world's most famous bear"—in previous years I could imagine some scientists thinking, demonstrated how? By what metrics?—cemented a spot in the scientific literature and natural history of the ecosystem. On May 16, 2023, wildlife watchers delighted to see the celebrated mother bear out of hibernation with a single new cub, in her traditional home range near Pilgrim Creek in Grand Teton National Park. Headlines announced the news that #399 was the oldest mama grizzly bear on record, at the age of twenty-seven.[12] In her twilight years, the matriarch maintained an arguably unmatched ability to amaze.

The bear's celebrity had prompted graduate student intern Lauren Sadowski—working with the Northern Rockies Conservation Cooperative in Jackson—to survey hunters, ranchers, agency personnel, animal advocates, and others in summer 2021 about their perceptions of the grizzly bear and their expectations for bear management. Results reflected varied values and beliefs. Hunters thought grizzlies should be delisted from Endangered Species Act protection to make it easier to control problem bears and reduce danger to hunters. Animal advocates and photographers felt wildlife managers didn't listen to them and didn't care for wildlife as much as they did. Ranchers felt unappreciated for how they protect habitat. Agency managers sought respect from people who thought that government workers liked having to kill bears. Bear biologists thought politics got in the way of good science.[13]

Both of the latter viewpoints resonate with me, although I have long felt that I belonged fully in neither the biologists' nor the managers' camp but on the periphery of both, a Forrest Gump–like witness to, and sometimes a participant in, historic happenings in the Greater Yellowstone Ecosystem. I lacked extensive wildlife research experience but had sufficient biological training to believe in managing at the population level; I have long resisted anthropomorphizing, or humanizing, individual subjects, as it can bias observations and complicate the tough decision-making that may be—that indeed is—required in conflict situations. And I never served as more than a temporarily "acting" deputy or park superintendent, though I sat with such leaders over thirty-plus years, presenting options and recommendations on how to react to bear encounters, property damages, and attacks. Managers at a high level typically approve (or, rarely, deny) biologists' recommendations on whether to trap, move, or remove an animal. Senior decision-makers are even more attuned to non-biological factors: community viewpoints, human safety risks, and the optics of actions taken. This is particularly true for something as visible as an endangered animal, and a world-renowned one at that. Federal and/or state park and wildlife leaders hold the responsibility for recovering grizzly bears and keeping people as safe as possible in bear country. I have yet to encounter a wildlife biologist or park manager who fell into their profession by accident or viewed it as "just a job, a paycheck" versus having a passion for nature, for conservation, in this case for bears. Their perceived steeliness in determining the fate of an individual animal and of a species' population should not, in my view, be mistaken for a lack of caring.

Wildlife managers in Sadowski's survey expressed discomfort over, if not dislike for, celebrity wildlife, as one might expect based on their training. Personally, the notion of "following" anyone, human or bruin, online or around a park is new and not appealing to me, not having grown up in the social media generation. Or I'm just of the slice of society that is ambivalent about celebrities in general. One summer day in the Tetons, rumors that one of the widely followed Kardashian sisters was in the park prompted some fans to excitement. I suspect that much of the visiting public went about their own pleasures unaware, nonchalant, or if involved in a positive encounter, momentarily captured by a photo opportunity. Perhaps it's this way with celebrity bears. Grizzlies being even more rare and far less knowable than human stars, a sighting—especially of a mother and cubs—can be the highlight of a Teton visit for many people. In the end, if no celebrity is made too uncomfortable by their audience, is there any harm?

The arguments against treating bears as luminaries include issues already discussed: they might become habituated to humans and be at higher risk of becoming food conditioned, a concern for both species. There is a robust market, too, for some wild animal pets (which is not permitted in Wyoming these days, though they are legal in some states), leading to incidents in which owners eventually release animals that get too large or too dangerous for home care into environments where they don't belong. People have strong views on whether it is appropriate to keep semi-domesticated wild animals at home, for entertainment, or in zoos. Other animal lovers can't help believing that wild creatures need and deserve human help, leading to casual or deliberate feeding. This has happened with bears and other wildlife in Jackson Hole—#399 and cubs from multiple litters were more than once attracted to molasses-enriched grain purportedly left out to feed moose, which is not illegal, on private property south of Grand Teton National Park.[14] Some people, in and outside agencies, dislike the notion that special treatment is given to celebrity animals. I have heard wildlife managers express the opinion that keeping bears like #399 around teaches the public the wrong lesson about what is still a risky proposition—that big predators and large numbers of humans can mutually and safely exist in a crowded semi-urban landscape like Jackson Hole. The viewpoint isn't heartless, nor is it aimed personally at *her*, so much as it reflects observed reality. Bear managers don't need much tenure to have been in the position to remove a bruin, and it is not pleasant to have to trap or otherwise put an animal down once a danger point has been crossed. Employees who make or carry out those decisions have experienced online harassment or worse from wildlife fans. After a mother grizzly killed a hiker in Yellowstone in 2015, park staff received death threats for killing the bear.[15]

Late in 2022, in Southern California, managers euthanized a twelve-year-old mountain lion, also with a number for a name, after he was hit by a car. Cougar P-22 had been caught by biologists working in the Santa Monica National Recreation Area and, like #399, he had become famous without knowing it, in films and books and online forums. Like a sports celebrity, he prompted lines of merchandise, from stuffed animals and pawprint casts to stickers and T-shirts, at least some of which provided proceeds to support a "Save LA Cougars" campaign. The big cat was memorialized in a sold-out celebration of his life held outdoors at Los Angeles's Greek Theatre, site of historic concerts by musicians Ringo Starr and Neil Diamond and, more recently, where scenes were filmed for the 2018 hit movie *A Star is Born*. Similar sentiment on a much smaller scale was shown in June 2016 when #399's sole cub of that year, named "Snowy" by local bear watchers, was killed by a vehicle strike near the Pilgrim Creek Road. Mourners created a small heart-shaped rock memorial, which was removed by park rangers as they closed the surrounding area to reduce attention and pressure on the adult bear.[16] After the passing of California's celebrity mountain lion, one commentator wrote that "the sentiment for P-22 rocketed past healthy human interest and wandered into serious obsession territory . . . anthropomorphizing wild animals has a long list of detriments, among them distorting our human relationship with the wild world around us . . . and creating a hierarchy of species worthy of concern that prioritizes the cute and fuzzy."[17]

My own discomfort with celebrity animals comes partly from my academic background but also from concern that in focusing on highly observable individuals, people may miss a broader story. Wildlife celebrities are such because they stand out among others of their kind—but they may not be unique so much as aided by a combination of chance and circumstances that put them in the right place at the right time to draw attention from fans. Certainly, their devotees should not obsess to the point of threatening wildlife managers who must decide the fate of, say, a grizzly bear that attacked a human being. I deeply hope that growing numbers of people will care about *all* wildlife, not just charismatic species or individuals that carry a moniker of celebrity. Bears and other creatures, like us, must have enough food, shelter, and security from whatever threatens them, which as often as not includes us humans. We can choose to offer stewardship, tempering how we live and recreate in their habitat, yielding enough for them to survive. And we do so understanding that neither cougars nor bears nor other wild creatures, iconic or not, care a whit for what we think of and do for them.

And yet.

P-22 lived a long life for a wild cougar. His fame arguably resulted in immeasurably increased understanding, affection, even tolerance for a predator that is not well known outside hunting circles and that has killed at least twenty-nine people in North America since 1890.[18] More tangibly, as widely shared before and after his memorial, awareness of the cougar's presence and his challenges living in the heavily trafficked Santa Monica Mountains prompted a successful campaign to build a $90 million wildlife crossing over U.S. Highway 101; ground-breaking occurred on Earth Day 2022. A wildlife bridge over the ten-lane Ventura freeway will reduce the dangers of roadkill for numerous species of animals while improving connectivity—genetic isolation is a special concern for the small local cougar population and other species. Is the value of that lessened because one evening six thousand people packed the Greek Theatre?

Sadly, bear #399 was killed by a car in the Snake River Canyon forty miles south of Jackson late in the evening of October 22, 2024. Her body was cremated and her ashes scattered near Pilgrim Creek in Grand Teton National Park. Her death inspired heartfelt tributes online, in print, and in person. Biologists gave her yearling cub, who disappeared, a good chance of surviving the winter on its own. Whether bears should be viewed in human terms or not, #399 was called a "good mother." She nurtured and protected enough cubs that some grew to raise their own, though the rate of her cubs' survival reflects a common challenge for bears: 53 percent of her offspring as of 2022 were known to have died, most as cubs or short of their reproductive adulthood, and it is hard to say whether habituation to humans contributed to this. Her fans call her special, which is likely not worth debating. Long before she set a record for cub-bearing at her advanced age, #399 was the first grizzly bear most people associated with Grand Teton National Park, where she demonstrated for the masses the adaptability of her species, her audiences, and the wildlife managers whose jurisdictions she has crossed.

And her "legacy": Bear warning signs at every trailhead, at picnic areas, even inside bathroom doors—a more cohesive, recognized bear management program for Grand Teton National Park and the Rockefeller Parkway. A permanent bear management specialist (with seasonal help) on the park staff. Grand Teton's Wildlife Brigade. A bear box in every campsite. One less cookout site in the willows, and a new fence around a sewage lagoon. Research on the relationship between bears and the elk reduction program, on visitor attitudes and knowledge about bears, and on the effectiveness of various signs and other educational approaches. The U.S. Fish and Wildlife Service hiring five bear-conflict specialists across greater Yellowstone, one stationed in Jackson Hole. Local newspaper headline writers and readers no longer assuming "Park" means only Yellowstone. And Teton County, Wyoming, at long last requiring bear-resistant trash containers throughout its jurisdiction.

Should we grant the term "legacy" to a bear who didn't care? Maybe not, but how many of these things, and others not listed here, would have happened without grizzly bear #399? Any of them? All of them? Should we have seen her, or some bear like her, coming? Probably. Isn't it the nature of bureaucracies and visitors and landowners and ranchers and nongovernmental organizations to be better at responding to than planning for change, to focus on what is in front of them today, to rely on routine in our subconscious expectations that things won't evolve, even when they continually do? No, she was not the first grizzly, even the first grizzly mother, to have lived in the Tetons, and she wasn't the last. But a combination of circumstances meant she was at the time and place that required human action, a much more heightened response than previously called for.

I do not think of her as an icon. (Breathe, readers, I mean her no insult.) It is celebratory enough to be a grizzly bear in greater Yellowstone or any ecosystem, to survive and even thrive. To have more than reproduced herself, as biologists hope for. To have provided countless hours of enjoyment and education, memories and artworks and photographs and video for millions of people around the world. To have engaged citizens in glimpsing an apex predator, in pondering and expressing viewpoints on whether grizzlies' existence in the ecosystem is threatened, on who should "manage" them, and on human relationships with wildlife, especially those that can kill us. If fans will temper their judgment of those with different views, we who are uncomfortable with celebrities can choose whether or how to seek them out; how others view a famous bear matters not to how I cherish Grand Teton Park or an animal itself. As a matter of history, bear #399 forced people to pay attention, to act, to rethink and reinforce how humans have responded to her and her kind in the Tetons—a story that will continue to play out across this and other places as wild animals show us what they may tolerate if we let them.

19

The Unknowable Bears

It was a pleasant mid-May evening drive around the park looking for wildlife and taking in the views. My visiting sister and I enjoy birdwatching, and we pulled off the road to watch a small hawk atop a tree just northeast of the Snake River Oxbow. It was not the first time we had bored my youngest daughter, just home from college for the year, who had not embraced birding but scanned the surrounding hillside instead. "Mom," she said, "I think I see a bear." I swung my binoculars uphill in the direction she indicated, replying, "Good spot. Wait—it has cubs. Of-the-year. Maybe," I half-joked, "it's #610." We watched the mother bear and her young move through the sagebrush, rather unusually all by ourselves for at least ten minutes before another car pulled over to see what we were watching. The more I looked, the more I thought it really might be the female grizzly who had been out of sight for a year and a half, to the point that there was, to me and my wildlife staff, rather too much speculation about her fate. Many wondered whether she had been killed, unreported, by a hunter or by another bear, or driven away from her Grand Teton home range by human harassment. I texted the bear management specialist that my party had a female with a pair of cubs-of-the-year in sight. I knew the park had no reports yet that season of a mother with two. Kate Wilmot immediately messaged me back that she had been chasing down such reports all evening, and over the next day or two staff became more certain that indeed we had confirmed the reappearance of grizzly #610. As promised, I put my daughter's name on the official bear sighting report to be sent to the Interagency Grizzly Bear Study Team.

Grizzly bear #610 was nearly as famous as her mother, #399, from whom she was born in 2006. By 2019 she had already given birth to three sets of her own offspring, beginning with her first litter in 2011, though a single cub seen in 2014 died early enough in the spring breeding season that #610 produced another group of cubs in 2015. This is not uncommon in bears. Early-season sightings of females with offspring can occur in April but are more likely in May. There was nothing surprising about the appearance of #610, save for the hype about her apparent absence since she last had cubs. She, like her even more well-known mother, commonly seemed to disappear from park roadsides when not tending precocious one- or two-year-old young. Observing her was special to me because of the company I kept and because it was my first grizzly sighting in the park that season—my last year working in Grand Teton.

To be honest, I had not worried about the bear or, as my career progressed, any other individual animal, even knowing that ones with a propensity for hanging too close to developments might have to be relocated or even removed. (Those occurrences caused a heavy sigh of resignation, which I fear my staff and my family heard too often.) I truly didn't fear the fate of bear #610, who had no history of human conflicts but had clearly demonstrated who would dominate a close encounter. So many bears have lived long lives out of our sight most if not all the time, and I admit I prefer it that way.

Though I enjoy scientific discoveries, I relish wild animals that, like old-time movie stars, retain an aura of mystery about their unseen lives. We can document where they go and what they eat and when they breed, and yet so many hours and miles of their lives are unknown to us. As long as their numbers and distribution and condition and habitat trend in positive directions, I have faith that they will remain long after I depart. I have faith in my fellow humans and even more in the animals themselves.

I can still picture the grizzled snout pushing on the front of the trap, the raspy tongue hanging an inch or two out of its mouth. The dull-ivory-colored claws reaching through the bars, and the small eyes, penetrating and steely. A deep huff echoed from the metal sides of the culvert trap holding a wild thing not at all happy to be confined. It was only for a moment, but it was still almost terrifying—"almost" only because the bars between us held.

During my career, I helped catch or handle some bruins but was never a bear trapper. Those are employees or members of the Interagency Grizzly Bear Study Team (IGBST), their primary mission to trap and radio collar a geographically distributed sample of grizzlies, particularly adult females, needed for monitoring the population throughout the ecosystem. Since the 1980s improvements in securing food, garbage, and other bear attractants have reduced the need to trap and relocate "management bears"—those that pose an increased risk to human safety by hanging around campgrounds or towns, by approaching people, or by regularly committing offenses police would equate to breaking and entering. As a result, park resource managers have considerably less need for, and thus practice in, handling bears each season—not that they lack the training and skills—and the assistance of the full-time bear trappers is valued when available.

I laughingly dated myself by the radio numbers of bears with which I had chanced to work. The study team numbers bears chronologically—grizzly bear #1 was trapped in May 1975, and in 2020 the number of grizzlies that had been captured and marked in the Greater Yellowstone Ecosystem topped a thousand. Today's staff and bear watchers are likely to be familiar with those radio-numbered in the higher hundreds or low thousands. Often, these are younger bears, caught due to inexperience making their own way in the wild world. They are less likely to be well established in a safe home range away from human eyes and temptations. The first grizzly bear I helped mark and relocate was #122, a three-year-old female caught in Yellowstone in April of 1986. Over the years, bears considered famous by the public, such as grizzly #134, who in her day was a celebrity at least in Yellowstone Park, have been atypical in their recognizability from living perilously close to areas frequented by humans. The great majority of bears—whether trapped and marked or not—are seldom seen by anyone. Their fates are known, if at all, only to pilots and researchers who count bears from the air, noting those that are collared among the greater number that are untouched, and to trappers who periodically catch and recapture them or retrieve a collar that emits a mortality signal from a dead animal. Twenty-first-century researchers estimate that 10–15 percent of the grizzly bear population in the ecosystem is caught and marked, meaning that most bears are never "known" to humans at all. I and every trapper and manager I have worked with like it that way.

My greatest frustration with moving bears from a site of the minor infraction of eating the right thing at the wrong place or time was that, often as not, we would not really know if or how the relocation was successful. If a bear wearing a radio collar later returned too close to human occupation, it was likely to be killed by managers, or by some member of the public defending a camp or even a life. If a marked bear disappeared, as hoped, into the wilderness, the collar would drop off or the battery would die within several years, and the bear might not be heard from or caught ever again. My older, retired self and trapper colleagues I've aged with recall tales of bears long gone, those we glimpsed but about whom we had mostly questions, wondering what happened to them once they broke the temporary bond of capture.

Two grizzly bear trappers dominated the research effort in my younger field days, during the 1980s and 1990s, and, as of 2024, they were both still actively working with bears, albeit in supervisory roles. Mark Haroldson is a calm, stocky man, dark-haired with a mustache, who speaks very softly while leaning close to your face. I suppose that trappers, if not predisposed to being quiet, get to be that way, just as they move efficiently without attention-grabbing motions, working around big predators most at home in the stillness and the dark. Beginning in 1991 I taught a grizzly bear class at the Yellowstone Institute, and early on I would ask Mark to stop by to tell attendees about bear trapping. For years he seemed uncomfortable inside the simple classroom, but he would lay out his trapping gear—snares, radio collars, and tracking antennae—among the sagebrush outside the historic Buffalo Ranch buildings and explain to students how and why he caught bears. His trapping tales, told with low-volume intensity and dry humor, made him the hit of the class. By the early 2000s Mark was hiring, training, and managing the seasonal trappers for the IGBST, having become my fellow instructor and a data guru, master of PowerPoint presentations, and coauthor on many publications referenced in this book about the ecosystem's grizzlies.

Mark's former trapping partner was Jamie Jonkel, a tall, lanky man with what has always sounded to me like a slight drawl. When I worked with him, he too spoke softly and deliberately after appearing to carefully contemplate what he wanted to say, which often came out with a chuckle. Jamie's dad, Charles "Chuck" Jonkel, was a wildlife biologist, too, renowned for studies of Canadian polar bears and of grizzlies in the Northern Rockies. When I first saw Chuck walking down the hall of the University of Montana Forestry building, where I went to graduate school, I knew it had to be the famous Dr. Jonkel—he *looked* like a bear, a little hunched over and stocky, with grizzled hair and beard through which his eyes appeared small and piercingly curious. I wondered if, as they say about some people and their pets, he chose to study bears because he resembled them, or if after a career watching them so closely he had come to assume a similar appearance. Jamie never came to remind me of a bear, despite years of trapping them and subsequently managing large carnivores around Missoula for the Montana Department of Fish, Wildlife and Parks. I doubt it stands out among his lifetime of bear-handling experiences, but Jamie was in charge during my most memorable instance of working with grizzlies, decades before I retired.

Bear #79 was an adult female well known to the early study team, born in 1974, first trapped in Montana, and moved to Yellowstone Park in 1981. What managers called a "good bear," she produced cubs with nearly predictable regularity every third year and seldom got into trouble raiding human foods or garbage. But in poor wild-food years, she found her way to the temptingly tasty apple trees found in various residential yards in Gardiner, Montana, just outside Yellowstone's North Entrance. In mid-August of 1990 the apples and some cooped-up chickens attracted #79 and her two yearling cubs into town during hyperphagia, the period of intense feeding when bears need to put on fat to sustain them through winter denning. To prevent her and the cubs from clashing with the good folks of Gardiner, biologists had set culvert traps, hoping to get the family group well out of town. Bear managers consulted as usual with the grizzly bear recovery coordinator and made a plan, in this case to separate the yearling cubs from their mother before she might instill in them potentially dangerous habits and to release them all into wilder areas away from the temptation of "city food." This meant putting as much distance as possible between temptation and the bears, while sending them to a friendly jurisdiction. (In those days, the practice of moving grizzlies across agency lines was not common and certainly took numerous phone calls.) Despite Yellowstone's and Grand Teton Park's different historical and managerial contexts, they are inextricably linked in ecology and mission, and I fondly recall their joint effort in this story.

The yearlings were destined for ground release in the John D. Rockefeller, Jr. National Memorial Parkway, and I was tapped, along with several others, to drive the yearlings, each in its own trap, the hundred miles across the length of Yellowstone. Park biologist Kerry Gunther would mount a helicopter slinging a trap carrying #79 to be released in that park's southeastern quadrant, in the Thorofare, the largest expanse of roadless wildland in the lower forty-eight states. The U.S. Forest Service prohibited, without regional forester approval, helicopter landings in the Teton and other congressionally designated wilderness areas abutting the parks. Sometimes we tried to run that administrative gauntlet, with varying success. Other times, we only half-joked that the park helicopter would hover over the flat plateau of Trident Mountain in Yellowstone, which has no official wilderness, and set a culvert down on the boundary with the trap door open, aimed toward the national forest, as we quietly urged, "Go, bear, go!"

Trapping situations are that strange combination of hurry-up-and-wait; anxiety to get the bear(s) out of harm's way as soon as possible is tempered by each animal's indifference to what we think. Trapping a family group is especially tricky, as inevitably each bear has its own idea about whether or when to get in a trap. This group had taken a few days to round up, but we wanted to release the year-and-a-half-old siblings together to increase their chances of surviving without the helpful watch of mom. Official rules called for not keeping grizzlies in traps for any longer than necessary, ideally less than twenty-four hours, which made us eager to get on the way to our respective locations. Our ground crews would face hours in the still-heavy visitor traffic, the helicopter could not fly after dark, and all of us were headed as far across Yellowstone as one could go.

Before the move could happen, though, #79 had to be in the right trap. She had been caught in a corrugated metal tube on wheels, eight feet long and three and a half feet around, likely scrounged from old stream culvert material, and weighing hundreds of pounds. An open mesh of two-inch squares covered the end opposite the solid trap door. Such open-ended traps had been common for years, but biologists had learned that bears often broke their teeth or claws on the metal squares. So as quickly as time and funds allowed, we replaced the old models with aluminum tubes, smooth-walled on both ends where the dangerous grates had been. These traps had small side windows that one could open from the outside to check on a bear's welfare or, when needed, to reach in with a jab-stick holding a syringe of an immobilizing drug to put the bear to sleep prior to placing a radio collar and ear tags on, or taking biological samples from, the animal. Yellowstone had invested in some of the newer smooth cylinders, which were also considerably lighter weight than the older culverts. Each had a welded metal attachment atop the center to facilitate the trap being hooked to a cable suspended beneath a helicopter. The challenge was to move #79 from the old steel culvert in which she sat to the newer, lighter aluminum trap for her ride to freedom.

We met at Kerry Gunther's rented park residence, a ramshackle historic two-story house outside of Gardiner. It was on a gravel "back road," along a tributary to the Yellowstone River. The administrative site provided plenty of open space and privacy from either accidental or intentional onlookers. The two female cubs, who under Mark Haroldson's direction had been newly collared and numbered #179 and #182, were held in their own aluminum tubes, while #79's trap was parked outside the long two-door wooden garage that, when needed, temporarily housed bears in relatively cool shelter from the surrounding stark, nearly treeless grass-and-sagebrush landscape.

Though Yellowstone Park resource biologists were in charge of management captures and relocations, on this occasion we gladly left the lead to study team trapper Jamie Jonkel. Kerry takes his work seriously but can appear slightly mischievous, quick to laugh or frown as befits the situation.

By 2024 the seasoned bear manager was approaching retirement age with countless bear handlings and nearly a hundred publications on his vita. In 1990 he was only a few years out of graduate school and not long into his permanent job as a bear management specialist. He had been around some bears for sure, but Jamie had experienced this rodeo before—a savvy adult bear #79, probably none too happy to have been caught, and needing to be moved from an old culvert. Jonkel told us how, in that prior instance, he had connected two traps so he could move her to a smooth-walled smaller trap, but the bear had other ideas. She sized up her temporary confinement cell until she found just enough purchase for a claw then ripped the traps apart and escaped. It's hard to picture in the abstract, even thinking you know how strong bears can be.

This time, we watched as Jamie backed the lighter aluminum trap up against the closed door of the culvert holding the mature mother grizzly. Under his strict direction, we jacked both traps up and down on various combinations of wooden and concrete blocks found in Gunther's garage. Once the doors matched as closely as possible, Jamie opened the newer cylinder and used large C-clamps to hold them together in multiple spots. He would not be rushed, repeatedly circling and eyeing the traps intently, periodically moving close to adjust and tighten the fasteners. The whole time, #79 sat quietly in her cage, barely moving. After what seemed like hours in the late summer sun, Jonkel was finally satisfied with his efforts to secure the two traps end to end. He stationed us carefully as well. A young seasonal employee, Nathan Varley, held a long rope tied to the open door of the aluminum trap, with strict instructions to drop it the moment he was signaled. Kerry and I were told to stand, silent and still, positioned off to the sides of #79's trap where she could not see us. Jamie raised the door of the old culvert, and we waited motionless for a while, but the bear didn't budge. She sat unmoving as she had for hours, quiet in the smelly culvert, wet with her drinking water and urine and remnants of the bait that had tempted her in.

Jamie looked at his watch and said softly, "We have to get going." Without warning, he grabbed a long, sturdy stick from the bed of his pickup truck and walked quickly back and forth along the length of the bear's trap, banging on it and yelling as loudly as I'd never heard him, "MOVE BEAR, MOVE!" In an equally abrupt response, #79 bolted at least twice to the far end of the aluminum trap then back to the near end of the old culvert, where she rammed her head and claws into the bars of the grate, anxious to find her escape. Next to me, Kerry Gunther stared wide-eyed, as I'm sure I was, transfixed by the speed and power of the grizzly as she showed her pure strength and urge to break out. Within seconds, Jamie yelled "NOW!" and Nathan dropped his rope, trapping the bear in the newer, cleaner trap. As suddenly as the commotion had started, it was over. Jamie turned to us and said, softvoiced again but insistent, "If you want a bear to do what you want it to do, you've got to be *aggressive*, you've got to be *dominant*."

My heart was still beating fast. My mental picture of that bear's sudden flip from a position of silently waiting to a sheer display of wild capability remains as vivid today as it was then. It has come first to my mind whenever I have heard someone say, "Oh, that bear wouldn't hurt anyone. I *know* her." Bear #79 was, until her end, not known to be aggressive to people, but the mostly unseen ability of a bear to defend, or attack, or both whenever it perceives a need should never be underestimated. While that certainly engenders respect or awe for bears, if not a healthy degree of fear, in me, the most amazing thing is that they so seldom unleash their power with full force upon us.

As #79 lay unmoving again in her new confines, we returned to the mundane details of our operation, freeing traps from now-unneeded clamps and blocks, re-hitching them to pickup trucks, and checking signals from radio collars. We grabbed our gear and, armed with snacks for the rest of the long workday, started driving. Kerry, aluminum trap in the back of one truck, headed to meet the helicopter pilot; Jamie and I drove separate trucks hauling the yearling grizzlies. Hours later, we met up with Grand Teton biologist Steve Cain near the (now removed) old Flagg Ranch cabins along the Snake River several miles south of Yellowstone's border. He led our small procession, late in the day, up the bumpy dirt-and-gravel Grassy Lake Road past Polecat Hot Springs to Glade Creek, where we backed the two traps up against the lodgepole forest, opened the doors, and awaited their departure for the freedom of the parkway and parts beyond. Once both young bears were free, we had an equally long ride home to Mammoth after a most unforgettable "day at the office."

That was thirteen years before I transferred to work for Grand Teton National Park and the Rockefeller Parkway, a time when visitors and staff seldom reported seeing grizzlies in either, and a time when Yellowstone had only recently begun asking for, and receiving, permission to relocate grizzlies outside its own jurisdiction. Back then managers weren't certain enough that transporting bears meant they might stay close to their release sites and out of trouble to merit the moves. As I made my way back north, I thought about the yearlings. Biologists judged that at the age of one and a half, they had only a 50 percent chance of making it to adulthood without their mother's guidance and care; female grizzlies typically keep their cubs for a second winter. Still, we hoped the effort was worth it and crossed our fingers that at least we had bought time for those bears, and who knew what time would bring. Though all three wore radio collars, those often wouldn't last more than two or three years on an adult female and even less time on young bears in growth mode. We knew that the fates of mother and cubs might remain unknown even to us bear movers.

Grizzly bear #79, freed in southeast Yellowstone that day, returned, as expected, by the next year to her home range in the northern part of that park. She lived to be twenty-two—fairly old for a wild mother bear—and had at least five litters, a minimum of ten cubs, before she was killed when she surprised an elk hunter north of Gardiner in the fall of 1996. The fate of female #182 is not well known, although in 1994 she was captured again by research trappers, at a backcountry site off the Flat Mountain Arm of Yellowstone Lake, and was reported as known to be alive in 1996.[1] She may have had one litter of cubs, whose survival is also unknown.[2]

I never saw bear #179 again after she disappeared into the woods at dusk that August 1990 day. But one afternoon, not long before he retired in February 2015, Steve Cain, who by then had worked with me at Grand Teton National Park for more than a decade, asked me if I remembered those bears from Yellowstone that we had released together in the parkway years before.

FIG. 20. Grizzly bear disappearing into forest. National Park Service photo.

As it turned out, #179 was a significant part of an analysis by the Grizzly Bear Study Team of "What to Do with the Offspring of Conflict Bears."[3] When the ecosystem's grizzly population was trending downward, managers considered alternate views on relocations. The first was that dependent offspring should be removed along with their mothers involved in conflicts.[4] The other idea was to separate and move yearlings to give them a chance to reach adulthood and recruit into the population without, perhaps, having learned risky behaviors from their parent. The young grizzlies we had moved in 1990 were, unknown to me, part of an experiment to test the latter strategy. Of fifty-three yearling grizzly bears translocated between 1981 and 2013, thirty-eight survived to the reproductive age of five. Shortly after her release, #179 spent time in the Owl Creek–Webb Canyon drainages in the northwest part of the Teton Range. After her first capture, she was trapped by researchers seven times between 1995 and 2012 and lived to be twenty-three or twenty-four years old. She was one of a dozen of the translocated yearlings known to have offspring of their own, mothering eleven to thirteen cubs in five or six litters, with only one known problem: getting into alfalfa pellets at a dude ranch along the Buffalo Fork of the Snake River. Her home range remained near her 1990 release site and to the east, and she and her offspring contributed significantly to reoccupation of grizzlies south of Yellowstone in and beyond Grand Teton National Park.[5]

Though I had heard nothing more of grizzly bear #179 during her two-plus decades of life, I still smile thinking of that successful management relocation, of a young female bear who went off alone and grew up in the wilds of the Tetons and matured to be a matriarch in her own right, without fanfare or public acclaim. She was neither a renowned nor a roadside bear; few people saw her. There have been and always are hundreds of those bears out there, occasionally found by a researcher or seen by a hiker or angler or elk hunter. They roam the moth sites along the eastern heights of the Absaroka Mountains, or whitebark pine stands and spawning streams and riparian zones and berry patches and the occasional livestock grazing allotments. They learn to live among their fellow bears and amid a remarkable buffet of food to sustain them, and they experience interactions we seldom see and can only imagine. Most of them avoid or ignore us. They are the vast majority of bears—both black and grizzly—in Grand Teton National Park and indeed across greater Yellowstone, unknown and always somewhat unknowable wild creatures that help make the ecosystem so special. May it always be so.

Notes

1. A BEAR BEGINNING

1. Schwartz, Miller, and Haroldson, "Grizzly Bear," 556–86.
2. Pyare et al., "Carnivore Re-colonisation," 71–77.
3. Fisher Smith, *Engineering Eden*, 39, 66.
4. White, Gunther, and van Manen, *Yellowstone Grizzly Bears*.
5. Mangelsen, *Grizzly 399*, 14.

2. WHERE BEARS BELONG

1. Sanborn, *The Grand Tetons*, 47, 57, 197.
2. Rockwell, *Giving Voice to Bear*, 97, 110.
3. Robert Stuart, *The Discovery of the Oregon Trail: Robert Stuart's Narratives of His Overland Trip Eastward from Astoria in 1812–13*, ed. Phillip Ashton Rollins (New York: C. Scribner's, 1935), 289, quoted in Whittlesey and Bone, *The History of Mammals in the Greater Yellowstone Ecosystem*, 30.
4. Schullery, *Lewis and Clark Among the Grizzlies*, 34.
5. Warren A. Ferris, *Life in the Rocky Mountains 1830–1845* (Denver: Old West Publishing Company, 1940), 163, quoted in Whittlesey and Bone, *The History of Mammals in the Greater Yellowstone Ecosystem*, 52.
6. Zenas Leonard, *Adventures of Zenas Leonard, Fur Trader and Trapper, 1831–1836*, ed. W. F. Wagner (1904), 249–51, quoted in Whittlesey and Bone, *The History of Mammals in the Greater Yellowstone Ecosystem*, 60.
7. Richard Leigh, "Writings of Richard 'Beaver Dick' Leigh of Jackson Hole, Wyoming," ed. Flora Edeen, Richard Leigh Manuscript Collection, University of Wyoming Archives, quoted in Saylor, *Jackson Hole, Wyoming*, 116.
8. Bonney and Bonney, *The Grand Controversy*, 370.
9. Nathaniel P. Langford, "The Ascent of Mount Hayden," *Scribner's Monthly*, June 6, 1873, 138, quoted in Whittlesey and Bone, *The History of Mammals in the Greater Yellowstone Ecosystem*, 252.
10. Sanborn, *The Grand Tetons*, 186–88.
11. William Henry Jackson, *The Pioneer Photographer* (New York: World Book Com-

pany, 1929), 126–32, quoted in Whittlesey and Bone, *The History of Mammals in the Greater Yellowstone Ecosystem*, 262–63.

12. Joseph E. Mushbach, Diary (1878), 7–8, original and typescript at Yellowstone National Park library, manuscript file. Also published in the (Helena) *Montana Record Herald*, December 1, 1929; and the *Billings* (MT) *Gazette*, August 16 and 23, 1936. Quoted in Whittlesey and Bone, *The History of Mammals in the Greater Yellowstone Ecosystem*, 531.

13. Baillie-Grohman, "Camps in the Rockies," 217–22.

14. Daugherty, *A Place Called Jackson Hole*, 90.

15. Judge, "Vital Laughter," 50.

16. Sanborn, *The Grand Tetons*, 244.

17. Daugherty, *A Place Called Jackson Hole*, 98.

18. News item, no title, *Jackson's Hole Courier* 6, no. 39 (September 24, 1914): 1.

19. "A Bear Christmas," *Jackson's Hole Courier* 6, no. 1 (December 31, 1914): 1.

20. "Young Lad Gets Big Game," *Jackson's Hole Courier* 28, no. 10 (July 4, 1918): 1.

21. "Bear Hunter Treed?" *Jackson's Hole Courier* 13, no. 42 (October 13, 1921): 1; and follow-up news item, no title, *Jackson's Hole Courier* 13, no. 43 (October 20, 1921), 1.

22. Daugherty, *A Place Called Jackson Hole*, 207–8.

23. Ad: "Bear for Sale," *Jackson's Hole Courier* 12, no. 27 (June 24, 1920): 6.

24. Huyler, *And That's the Way it Was in Jackson Hole*, 9–10.

25. Huyler, *And That's the Way it Was in Jackson Hole*, 212.

26. "Grizzly Bear from Asia," *Jackson's Hole Courier*, 11, no. 25 (June 12, 1919): 2.

27. Untitled news item on meeting of Wind River Livestock Growers' Association, *Jackson's Hole Courier*, 13, no. 31 (July 21, 1921): 4.

28. Blair, *The History of Wildlife Management in Wyoming*, 70.

3. BEAR MANAGEMENT

1. Daugherty, *A Place Called Jackson Hole*.

2. Righter, *Crucible for Conservation*, 8–11, 20–41.

3. U.S. Department of the Interior, National Park Service, *Report of the Director 1929*, 2, 85.

4. Hanks, "Talus," 35.

5. Wondrak Biel, *Do (Not) Feed the Bears*, 36–37.

6. Bennett Gale for Superintendent Smith, *Reports of the Naturalist Division*, August 2 and November 1, 1940, NARA-KC, Record Group 79, Box 77, File 207.04.

7. Smith, Chas. J. to director, *Annual Wildlife Report for Grand Teton National Park*, October 3, 1940, NARA-KC, Record Group 79, Box 77, File 207.01.

8. Blair, *The History of Wildlife Management in Wyoming*, 41, 58.

9. Franke, Paul R., *Annual Wildlife Report*, October 3, 1944, NARA-KC, Record Group 79, Box 77, File 207.01.

10. "State Animals, Bird and Fish Preserves, and Blue Holes State Park," 48.

11. "U.S. Judge Denies Federal Motion in Monument Suit," *Cheyenne Tribune*, reprinted in the *Jackson's Hole Courier* 26, no. 23 (December 2, 1943): 1.

12. Blair, *The History of Wildlife Management in Wyoming*, 113–14.

13. Righter, *Crucible for Conservation*, 120.

14. Blair, *The History of Wildlife Management in Wyoming*, 117–18.

15. Allyn F. Hanks, *Annual Park Wildlife Report*, n.d., ca. 1945–46, NARA-KC, Record Group 79, Box 77, File 207.01.

16. Jepson, Carl R. to superintendent, containing the *Park Naturalist's Report for July 1948*, August 9, 1948, NARA-KC, Record Group 79, Box 78, File 297.04.

17. Pyare et al., "Carnivore Re-colonisation," 74.

18. Franklin K. Lane, "Letter to Mather on National Park Management," May 13, 1918, in Albright and Cahn, *Birth of the National Park Service*, 69.

4. BOOM TIMES

1. Ziesler and Spalding, *National Park Service Visitor Use Statistics*.

2. Blair, *The History of Wildlife Management in Wyoming*, 152–53.

3. District ranger (initials indecipherable) Moose District, to chief ranger, *Bear Incident Report for 1959*, undated, GTNP, Accession 530 (Ursids), Series 10.01, Box 1, File Unit 2.

4. Acting Superintendent Yeager to regional director, Region Two, *Bear Incidents Report 1960*, November 29, 1960, GTNP, Ursids, Series 10.01, Box 1, File Unit 2.01.

5. Acting Director Tolson to Washington office and all field offices, *National Park Service Bear Management Program and Guidelines*, July 6, 1960, GTNP, Ursids, Series 10.01, Box 1, File Unit 3.

6. Acting Superintendent Peterson Jr. to regional director, Midwest Region, *Bear Management Activities Report, 1965*, December 3, 1965, GTNP, Ursids, Series 10.01, Box 1, File Unit 2.01.

7. Acting Superintendent Mitchell to regional director, *Bear Management Activities Report, 1966*, December 8, 1966, GTNP, Ursids, Series 10.01, Box 1, File Unit 2.

8. "Camp Robber Bear Helps Himself to 15 Cans Beer," *Jackson Hole Guide*, August 5, 1965, 15.

9. "Bears Are Hungry," *Jackson Hole Guide*, October 6, 1966, 8.

10. "Rural School News, Moran School News," *Jackson Hole Guide*, October 11, 1962, 7.

11. "Dr. Craighead to Show Grizzly Bear Film in Jackson," *Jackson Hole Guide*, April 6, 1967, 5.

12. "Byrle Walker Mauled by Grizzly," *Jackson Hole Guide*, June 8, 1967, 1.

13. "Grizzly Bears Kill Two Girls in Glacier Park," *Jackson Hole Guide*, August 17, 1967, 6; and "Safety in the Wilderness," *Jackson Hole Guide*, August 24, 1967, 2.

14. "Editorially Speaking . . . ," *Jackson Hole Guide*, August 1, 1968, 3.

15. "Black Bears Work on 422, Grizzlies 20 in 8 Years," *Jackson Hole Guide*, August 22,

1968, 1; and "Grizzly Zoo Hardly an Answer," *Jackson Hole Guide*, August 22, 1968, 5.

16. Sellars, *Preserving Nature in the National Parks*, 201, 243–44.

17. *Bear Management Plan Grand Teton National Park—1970*, found with memorandum from Superintendent Kerr to Alan Atchison, chairman; Bob Wood, Bill Barmore, Dickie Stansbury, and Joe Shellenberger, *1977 Bear Management Program*, GTNP, Ursids, Series 10.01, Box 1, File Unit 4.

18. Sellars, *Preserving Nature in the National Parks*, 239.

19. Sellars, *Preserving Nature in the National Parks*, 204–66.

20. *Wildlife Observations*, 1970 and 1973, GTNP, Ursids, Series 10.04, Box 1, File Unit 8.

21. Superintendent to director, Midwest Region, *Bear Management Activities Report—1971*, December 20, 1971, GTNP, Ursids, Series 10.01, Box 1, File Unit 2.

22. Schullery, *The Bears of Yellowstone*, 144.

23. U.S. Department of the Interior, Fish and Wildlife Service, *Amendment Listing the Grizzly Bear of the 48 Conterminous States as a Threatened Species*.

24. Robert P. Wood, *Survey of Current Bear Problems and Related Management Activities in the National Park System, Responses for Grand Teton National Park (1975) and the JDR Parkway (1977)*, GRTE, Ursids, Series 10, Box 1, File Unit 4 and 2_02.

25. Chief of park maintenance Caresia to superintendent, *Solid Waste Program*, June 10, 1975, GTNP, Ursids, Series 10.01, Box 1, File Unit 4.

26. "Meeting Local Fauna, Park Inaugurates Animal Identification Course for Rangers," *Jackson Hole News*, September 29, 1976, 19/B3.

27. Superintendent Kerr to Alan Atchison, chairman; Bob Wood, Bill Barmore, Dickie Stansbury, and Joe Shellenberger, *1977 Bear Management Program*, October 13, 1976, GTNP, Ursids, Series 10.01, Box 1, File Unit 4.

28. *Grand Teton National Park Bear Management Plan 1977*; and *Grand Teton National Park* [and] *John D. Rockefeller, Jr. Memorial Parkway Bear Management Policy*, 1977, 2, GTNP, Ursids, Series 10.01, Box 1, File Unit 4.

29. *Bear Management Summary 1977, Grand Teton National Park, John D. Rockefeller, Jr. Memorial Parkway*, GTNP, Ursids, Series 10.01, Box 1, File Unit 2.02.

5. MAKING THE LIST

1. "A Look Into the 80s, Public Lands," *Jackson Hole News*, January 3, 1980, 17.

2. Hoak, Clark, and Wood, "Grizzly Bear Distribution," 245–47.

3. "Is Critical Habitat People vs. Bears?" *Jackson Hole News*, November 10, 1976, 1, 3.

4. "Grizzly Controversy Stirs Local Fears," *Jackson Hole News*, November 17, 1976, 1.

5. Craighead, "A Proposed Delineation of Critical Grizzly Bear Habitat," 17.

6. Sellars, "The Grizzly State of the Endangered Species Act," 475, 499–500.

7. U.S. Department of the Interior, U.S. Fish and Wildlife Service, *Grizzly Bear Recovery Plan*, 1982, 39–44.

8. Mealey, *Interagency Grizzly Bear Guidelines*, 3–4.

9. Robert F. Wood, *Bear Incident Management Action Report 1984*, GTNP, Ursids, Series 10.01, Box 1, File Unit 2.01.

10. Katy Duffy to Bob Wood, *Bearproof Garbage Containers*, August 8, 1984, GTNP, Ursids, Series 10.01, Box 2, File Unit 19.

11. Robert F. Wood, *Bear Incident Management Action Report 1986*, GTNP, Ursids, Series 10.01, Box 1, File Unit 2.02.

12. A. Robb, *Bear Incident and Management Action Report, 1988*, February 20, 1989, GTNP, Ursids, Series 10.01, Box 1, File Unit 2.02.

13. Grand Teton National Park, *1989 Bear Sightings and Incident Statistics for Grand Teton National Park and John D. Rockefeller, Jr. Memorial Parkway*, no date, GTNP, Ursids, Series 10.01, Box 1, File Unit 2.02.

14. Todd Wilkinson, "Grizzly 139: A Bad Bear Gone Straight?" *Jackson Hole News*, June 29, 1988, 38.

15. Resource Biologist Mark T. Schroeder to chief, Maintenance Division, trash collection facilities, July 28, 1987, GTNP, Ursids, Series 10.01, Box 2, File Unit 19.

16. Harold W. Werner to superintendent and squad, "Food Locker Needs," October 24, 1989, GTNP, Ursids, Series 10.01, Box 2, File Unit 18.

17. Assistant Chief Ranger Edward H. Christian to assistant superintendent, operations, through chief park ranger, *Bearproofing*, April 11, 1990, GTNP, Ursids, Series 10.01, Box 2, File Unit 19.

18. Wildlife Biologist Steven L. Cain to superintendent, assistant superintendent, and others, *Minutes of Bearproofing Meetings*, December 20, 1990, and December 19, 1991; the latter with a handwritten note initialed by Superintendent Jack Stark complimenting Cain, as quoted in the text; GTNP, Ursids, Series 10.01, Box 2, File Unit 19.

19. North District ranger Don Coelho to chief park ranger, *Bearproof Dumpsters at Pilgrim and Pacific Creek Hunt Camps*, October 3, 1992, GTNP, Ursids, Series 10.01, Box 4, File Unit 36.

20. "Table of Bear Conflict Sites by Agency," IGBC-YES, undated report, ca. 1990.

21. "Grizzly Coordinator Predicts Recovery with Next Decade," *Jackson Hole News*, March 29, 1989, 9.

6. THE LIFE OF A BEAR

1. Pelton, "Black Bear," 552; Schwartz, Miller, and Haroldson, "Grizzly Bear," 571.

2. Schwartz, Haroldson, and Cherry, "Reproductive Performance of Grizzly Bears," 19–22.

3. Mike Koshmrl, "A Record Brood of Baby Bruins," *WyoFile*, November 3, 2023.

4. Robbins et al., "Maternal Condition Determines Birth Date and Condition of Newborn Bear Cubs," 543.

5. Haroldson et al., "Grizzly Bear Denning Chronology and Movements," 32–33.

6. Holm, Lindzey, and Moody, "Interactions of Sympatric Black and Grizzly Bears," 102.

7. Frattaroli, "Black Bear Ecology," 78.
8. White, Gunther, and van Manen, *Yellowstone Grizzly Bears*, 66.
9. White, Gunther, and van Manen, *Yellowstone Grizzly Bears*, 67.
10. Schullery, *Lewis and Clark among the Grizzlies*.
11. Holm, Lindzey, and Moody, "Interactions of Sympatric Black and Grizzly Bears," 103–5.

7. DUMP DAYS

1. Johnson, Daily Diary Entry.
2. Wondrak Biel, *Do (Not) Feed the Bears*, 19–20, 66, 78–79, 95.
3. "Valley Pioneer Maude Timmermeyer Remembered," *Jackson Hole News*, December 8, 1993, 43/11B.
4. Johnson, Daily Diary Entry.
5. Emails from Larry Rockefeller to author, February 13, 2021, and November 30, 2022. Used with permission.
6. Email from Steven Rockefeller, shared via email from Larry Rockefeller to author, February 18, 2021. Used with permission.
7. Wood, *Bear Incident Management Action Report 1986*, GTNP, Ursids, Series 10.01, Box 1, File Unit 2.02.
8. Huyler, *And That's the Way It Was in Jackson's Hole*, 242–43.
9. Conversations with Judith Allyn Schmitt, October 11 and November 3, 2021. Used with permission.
10. Emails from Karin Gottlieb to author, September 9, 2021. Used with permission.
11. Conversation with Fred Herbel, September 13, 2021. Used with permission.
12. Conversation with Cynthia Galey Peck, November 8, 2021. Used with permission.
13. "Grand Teton National Park, Permit for Disposal of Household Waste in Sanitary Landfills within Grand Teton National Park," undated, signed by Irwin Lesher for X Quarter Circle X Ranch, JHHSM, Item #2002.118.63.
14. Jim Miller, "Snake Management Plan May Limit Float Trips," *Jackson Hole Guide*, July 18, 1974, 3.
15. Conversation with and email from Steve Baldock, Grand Teton National Park maintenance employee, November 23, 2020.
16. Yeager to regional director, *Bear Damage—Personal Injury and Property Damage for 1958*, February 2, 1959, GTNP, Ursids, Series 10.01, Box 2, File Unit 1.
17. Fagergren to regional director, Midwest Region, *Bear Management Activities Report for 1964*, November 27, 1964; Peterson to regional director, *Report for 1965*, December 3, 1965; Mitchell to regional director, *Report for 1966*, December 8, 1966; Chapman to regional director, *Report for 1967*, December 4, 1967; and *Report for 1968*, December 3, 1968, GTNP, Ursids, Series 10.01, Box 2, File Unit 1.
18. Young to regional director, *Bear Management Activities Report—1969*, December 4, 1969, GTNP, Ursids, Series 10.01, Box 2, File Unit 1.

19. Conversation with and email from Steve Baldock, Grand Teton National Park maintenance employee, November 23, 2020. Used with permission.
20. Conversation with John McAvoy, retired Grand Teton National Park maintenance employee, August 23, 2021. Used with permission.
21. Conversation with Barry Alexander, retired Grand Teton National Park maintenance employee, August 23, 2021. Used with permission.
22. Conversation with Steve Cain, retired Grand Teton National Park senior wildlife biologist, September 1, 2012, Jackson, Wyoming. Used with permission.
23. "County Commissioners Meet," *Jackson Hole News*, August 5, 1971, 9.
24. "Joint City-County Meet," *Jackson Hole News*, November 4, 1971, 26.
25. "Keep Teton County 'Out of the Dumps,'" *Jackson Hole Guide*, August 17, 1972, 8.
26. Thuermer, Angus M. Jr., "Garbage-to-Energy Method Eyed," *Jackson Hole News*, March 23, 1983, 11.
27. Conversation with Becky Kiefer, Teton County Integrated Solid Waste Management Department, November 22, 2021.
28. Judy Hockelberg, "Moran News," *Jackson Hole News*, October 1, 1970, 3.

8. BERRIES AND MORE

1. Muir, "Among the Animals of the Yosemite," 617.
2. Gunther et al., "Dietary Breadth of Grizzly Bears," 61.
3. Costello et al., "Diet and Macronutrient Optimization in Wild Ursids," 8.
4. Frattaroli, "Black Bear Ecology," 24–26, 29–32, 123–26.
5. "Bears Close Campground Early," *Jackson Hole News and Guide*, October 14, 1976, 3.
6. "Scientist Spotlight: Whitebark Conservation," in *Natural and Cultural Resources Vital Signs 2021*, Grand Teton National Park and John D. Rockefeller, Jr. Memorial Parkway, 14–15; and "Whitebark Pine," in *Natural and Cultural Resources Vital Signs 2021*, Grand Teton National Park and John D. Rockefeller, Jr. Memorial Parkway, 32.
7. Personal communication with Dan Reinhart, retired resource manager, Grand Teton National Park, March 2024, and Jennifer Whipple, retired botanist and herbarium curator, Yellowstone National Park, June 2022. Used with permission.
8. Frattaroli, "Black Bear Ecology," 124.
9. Costello et al., "Diet Optimization," 16–17.
10. O'Ney and Gipson, "A Century of Fisheries Management in Grand Teton National Park," 132.
11. Email from Chad Whaley, fisheries biologist at Grand Teton National Park, April 27, 2021.
12. "Another Roadside Attraction: Grizzly No. 399 and Family Find Fish Caches at Oxbow Bend to the Delight of Dozens of Observers," *Jackson Hole News and Guide*, May 7, 2008, 17A.
13. Smith and Anderson, "Patterns of Neonatal Mortality of Elk," 1233–34.
14. Smith et al., "Neonatal Mortality of Elk," 162.

15. Berger, "Fear, Human Shields, and the Redistribution of Prey and Predators," 622.
16. Conversation with Craig Whitman, Interagency Grizzly Bear Study Team, 2022.

9. BEARS AND LIVESTOCK, PART 1

1. Consolo-Murphy, "Holding on to Yellowstone's Grizzlies," 4–5.
2. Miller and Hardy, *Final Project Report*, 1.
3. Paul R. Franke, *Grazing of Domestic Stock, Report for Calendar Year Ending December 31, 1945*, January 17, 1946, NARA-KC, Box 77, Folder 207.01.
4. "Mid-winter Bear Story," *Jackson Hole Guide*, March 1, 1962, 1.
5. Conversation with Cynthia Galey Peck, November 8, 2021. Used with permission.
6. Daugherty, *A Place Called Jackson Hole*, 89–97.
7. "Teton National Forest Will Mark Fifty-Second Anniversary February 22," *Jackson's Hole Courier* 21, no. 33 (February 17, 1949): 3.
8. "How Wyoming Representatives First Favored Park Extension," *Jackson's Hole Courier*, October 27, 1932, 1.
9. Righter, *Crucible for Conservation*, 29–30.
10. Daugherty, *A Place Called Jackson Hole*, 155.
11. Franke, Paul R. *Grazing of Domestic Stock, Report for Calendar Year Ending December 31, 1945*, January 17, 1946, NARA-KC, Box 77, Folder 207.01.
12. Daugherty, *A Place Called Jackson Hole*, 155.
13. Johnson and Griffel, "Sheep Losses on Grizzly Bear Range," 786–90.
14. Jorgensen, "Bear-Livestock Interactions," 76–91, 132.
15. Barker, *Saving All the Parts*, 43–44.
16. Knight and Judd, "Grizzly Bears That Kill Livestock," in *Yellowstone Grizzly Bear Investigations*, 46–48.
17. Conversation with Dan Tyers, June 1, 1922, Bozeman MT.
18. Knight and Judd, "Grizzly Bears That Kill Livestock," in *Bears: Their Biology and Management*, 189. This article omitted one of the quoted lines in the text. The full quotation can be found in Knight and Judd, "Grizzly Bears That Kill Livestock," *Yellowstone Grizzly Bear Investigations*.
19. Blanchard, Knight, and Yount, "Mortality," 20.
20. Knight, Blanchard, and Kendall, "Recommendations and Conclusions," in *Yellowstone Grizzly Bear Investigations, 1980*, 46.
21. Allen et. al., "Report to Secretary of the Interior," 12–13.
22. Johnson and Griffel, "Sheep Losses on Grizzly Bear Range," 787.
23. William Nell, "Mutton-Loving Bear May Force Sheep to Move," *Billings Gazette*, September 10, 1983, 10-A.
24. "Last of the Sheep Leave Yellowstone Grizzly Country," *Billings Gazette*, October 4, 1988, 11-A.
25. McCrystie Adams, "Grizzlies Suspected of Sheep Kills on Targhee," *Jackson Hole Guide*, September 4, 1996, A10.

26. U.S. Department of the Interior, Fish and Wildlife Service, *Proposed Rule Designating the GYE Population of Grizzly Bears as a Distinct Population Segment*, 2005, 69867–69868.

27. U.S. Fish and Wildlife Service, *Biological Opinion for Caribou-Targhee National Forest*, 2016, 41.

10. BEARS AND LIVESTOCK, PART 2

1. NPS, Grand Teton NP. Interpretive wayside located on U.S. Hwy 89 pullout overlooking the Elk Ranch.

2. *An Act to Establish a New Grand Teton National Park*, 850–51.

3. Conversation with Mallory Smith, chief of business and administration for Grand Teton National Park, October 14, 2022.

4. Righter, *Peaks, Politics and Passion*, 94.

5. Turner, Oral Interview, August 1, 2017, 35–36.

6. Grand Teton National Park and John D. Rockefeller, Jr. Memorial Parkway, *Natural and Cultural Resources Vital Signs 2019*, 39.

7. Righter, *Peaks, Politics and Passion*, 82–83.

8. "Interview with Maynard Burrows, September 20, 1972," NPS Oral History Collection, hfca_1817_II_035_barrows_maynard, 36; and "Interview with Russell E. Dickinson, September 26, 1962," NPS Oral History Collection, 012_Dickenson, Russell E., 14.

9. Jepson, Carl E. *Monthly Report[s] of Park Naturalist, August and October 1951*, September 5 and November 2, 1951, NARA-KC, Record Group 79, Box 78, Folder 297.04.

10. Grand Teton National Park and John D. Rockefeller, Jr. Memorial Parkway, *Natural and Cultural Resources Vital Signs 2013*, 24.

11. Humstone, "Elk Ranch Determination of Eligibility," 42.

12. Knight and Judd, "Grizzly Bears That Kill Livestock," *Yellowstone Grizzly Bear Investigations*, 47.

13. "Two Grizzlies Guilty of Recent Cattle Raids," *Jackson's Hole Courier*, August 3, 1933, 1.

14. Murie, "Cattle on Grizzly Bear Range."

15. Blair, *The History of Wildlife Management in Wyoming*, 102.

16. Pete Hayden, chief, resource management, Grand Teton National Park to Norm Bishop, Yellowstone National Park, *Transmittal of Grazing Data* (with accompanying map), April 23, 1991, GTNP, Accession 474, Lands Records-Grazing, Series 003.01, Box 3, File Unit 17.

17. Righter, *Peaks, Politics and Passion*, 88.

18. Conversation with Steve Cain, retired Grand Teton National Park senior wildlife biologist, September 1, 2021, Jackson, Wyoming. Used with permission.

19. Russell, *Grizzly Country*, 52.

20. Brandon Loomis, "Grizzly Bear Plagues Walton Cattle," *Jackson Hole Guide*, Octo-

ber 14, 1992, 1, 13.

21. Todd Wilkinson, "Togwotee Griz May Be Moved," *Jackson Hole News*, October 6, 1993, 1, 23; and Angus M. Thuermer Jr., "Togwotee Griz Gets a Reprieve," *Jackson Hole News*, October 13, 1993, 1, 19.

22. Conversation with Mark Bruscino, retired Wyoming Game and Fish Department biologist, October 21, 2022, Cody, Wyoming. Used with permission.

23. Mealey, *Interagency Grizzly Bear Guidelines*, 3–4.

24. Wyoming Game and Fish Department biologist Dave Moody and grizzly bear recovery coordinator Christopher Servheen to Yellowstone Ecosystem Grizzly Bear Subcommittee chairman Robert D. Barbee, September 2, 1993, IGBC-YES.

25. Conversation with Mark Haroldson, IGBST biologist, June 1, 2022, Bozeman, Montana.

26. Anderson, Ternant, and Moody, "Grizzly Bear–Cattle Interactions on Two Grazing Allotments," 247–56.

27. Angus M. Thuermer Jr., "Plan to Move and Kill Cow-Eating Bears Hit," *Jackson Hole News*, May 3, 1995, 3A.

28. Gunther et. al., "Grizzly Bear–Human Conflicts 1994," 14, also Appendix A, 34.

29. Angus M Thuermer Jr., "Public Comment Sought on Walton Grazing Plan," *Jackson Hole News*, July 27, 1994, 9A.

30. *Case Incident Record 952653*, August 27, 1995, GTNP, Ursids, Series 10.4, Box 23, File Unit 10.

31. Gunther, et. al., "Grizzly Bear–Human Conflicts 1995," 14, also Appendix A, 27.

32. Angus M Thuermer Jr., "Greens Battle Plan to Move Raiding Griz," *Jackson Hole News*, September 20, 1995, 1, 23A.

33. Anderson, Ternant, and Moody, "Grizzly Bear–Cattle Interactions on Two Grazing Allotments," 247–56.

34. Angus M Thuermer Jr., "Traps Laid in Park for Cow-Eating Griz," *Jackson Hole News*, July 24, 1996, 1, 22A.

35. Webb, *A Woman in the Great Outdoors*, 188–90.

36. Conversations with Steve Cain, September 1, 2021, Jackson, Wyoming, and with Mark Bruscino, retired Wyoming Game and Fish Department biologist, October 21, 2022, Cody, Wyoming. Used with permission.

37. "IGBC Bear Management Action Consultation Form," August 13, 1996, GTNP, Ursids, Series 10.01, Box 5, File Unit 48.

38. McFetters and Cain, *Bear Sightings, Management Actions, and Educational Efforts, Grand Teton NP/John D. Rockefeller, Jr. Memorial Parkway*, October 1996, 4–5, GTNP, Ursids, Series 10.01, Box 2, File Unit 11.02.

39. Grand Teton National Park, "Grizzly #309 Captured in Grand Teton National Park," Press release issued by Linda Olson, August 4, 1996, GTNP, Ursids, Series 10.01, Box 5, File Unit 48.

40. Angus M. Thuermer, Jr., "New Plan Aims to Keep Cows, Grizzlies Separated," *Jackson Hole News*, April 16, 1997, 1, 19A.

41. Angus M. Thuermer Jr., "Grizzly No. 203 Targeted," *Jackson Hole News*, July 16, 1997, 1, 23; and "Bear 203 Runs from Guns, Firecrackers," *Jackson Hole News*, July 23, 1997, 3A.

42. Rachel Odell, "Where Is Bear No 203?" *Jackson Hole News*, December 2, 1998, 8A.

43. David Simpson, "Game and Fish Reworks Grizzly Loss Formula," *Jackson Hole Guide*, March 18, 1998, A10.

44. Renee Leshan and Steve Cain, *1999 Bear Management Program, Grand Teton National Park and the John D. Rockefeller, Jr. Memorial Parkway*, October 4, 1999, GTNP, Ursids, Series 10.01, Box 2, File Unit 17.

45. Rachel Odell, "Electric Fence Foes up between Griz, Cattle," *Jackson Hole News*, July 28, 1999, 32A.

46. Grand Teton National Park, *Biological Assessment, Threatened and Endangered Species for Reissuance of Grazing Permits*, 21.

47. Anderson, Ternant, and Moody, "Grizzly Bear–Cattle Interactions on Two Grazing Allotments," 254–55.

48. Rebecca Huntington, "Park Grazing Faces Uncertainty," *Jackson Hole Guide*, September 29, 1999, 1, A19.

49. Rachel Odell, "Ranching Veteran Gives Heartfelt Testimony," *Jackson Hole News*, November 10, 1999, 10B.

50. Grand Teton National Park, *Biological Assessment, Threatened and Endangered Species for Reissuance of Grazing Permits*, 31.

51. Rebecca Huntington, "New Outbreak Alters Brucellosis Picture," *Jackson Hole News and Guide*, August 4, 2004, 11A.

52. Mary Gibson Scott and Carole (Kniffy) Hamilton, superintendent, Grand Teton National Park, and supervisor, Bridger-Teton National Forest, to files, Pacific Creek Grazing Allotment, May 28, 2009, GTNP, Lands Records-Grazing, Series 003.02, Box 2, File Unit 6.7.

53. Mary Gibson Scott and Jacque Buchanan, superintendent, Grand Teton National Park, and supervisor, Bridger-Teton National Forest, to Ernie Cockrell, YY Partners, L.P., November 22, 2010, and response January 13, 2011, GTNP, Lands Records-Grazing, Series 003.02, Box 2, File Unit 6.7.

54. Mike Koshmrl, "A Chunk of Grand Teton Park Could Go up for Auction. Price Tag $62M," *WyoFile*, October 2, 2023; and "Wyoming Legislature's Two Chambers OK 'Kelly Parcel' Sale to Feds for $100M," *WyoFile*, February 22, 2024.

55. Grand Teton National Park and John D. Rockefeller, Jr. Memorial Parkway, *Natural and Cultural Resource Vital Signs 2019*, 39.

56. Conversation with Jason Wilmot, Bridger-Teton National Forest Blackrock district ranger, January 13, 2023.

11. WAKE-UP CALL FOR A NEW CENTURY

1. Schwartz et al., "Distribution of Grizzly Bears, 1990–2000," 202–13; and "Distribution of Grizzly Bears, 2004," 63–66.

12. BEAR DANGER

1. Stringham, Rogers, and Bryan, "Have Black and Grizzly Bears Become More Dangerous?" 2.
2. Herrero, *Bear Attacks*.
3. Herrero et al., "Fatal Attacks by American Black Bear," 598.
4. Interagency Grizzly Bear Study Team, *Bear-Caused Human Fatalities*, June 2023. Note: As of this writing, one additional fatality occurred outside West Yellowstone, Montana, after the table was posted.
5. "Pioneer Mother Is Called by Death," *Jackson Hole Guide*, December 20, 1954, 16.
6. Acting Superintendent Peterson Jr. to regional director, Midwest Region, *Bear Management Activities Report, 1965*, December 3, 1965, GTNP, Ursids, Series 10.01, Box 1, File Unit 2.01.
7. Superintendent Chapman to regional director, Midwest, *Bear Management Activities Report, 1967*, December 4, 1967, GTNP, Ursids, Series 10.01, Box 1, File Unit 2.01.
8. Superintendent Kerr to associate regional director, park system management, Rocky Mountain Region, *1976 Bear Incident and Management Action Report*, December 3, 1976, with enclosed report prepared by Robert Wood, GTNP, Ursids, Series 10.01, Box 1, File Unit 2.01.
9. "Bear Baiter Bitten," *Jackson Hole News*, August 11, 1976, 4.
10. "Bears Shut Jenny Lake Area," *Jackson Hole News*, October 13, 1976, 1.
11. Robert Wood, *Bear Management Summar[ies] 1977 and 1978, Grand Teton National Park [and] John D. Rockefeller, Jr. Memorial Parkway*, January 15, 1979, GTNP, Ursids, Series 10.01, Box 1, File Unit 2.02.
12. Robert Wood, *Black Bear Incident and Management Action Report 1979*, January 1980, GTNP, Ursids, Series 10.01, Box 1, File Unit 2.02.
13. David Hackett, "Community Mourns Susan Walker Death," *Jackson Hole Guide*, August 20, 1985, A3. Also, Angus M. Thuermer Jr., "Rangers Shed Light on Death of Susan Walker in Park," *Jackson Hole News*, August 21, 1985, 12.
14. Mary Beth Baptiste and Steve Cain, *Bear Sightings, Educational Efforts, and Management Actions, Grand Teton National Park 1994*, October 1994, 7, GTNP, Ursids, Series 10, File Unit 11.01.
15. David Simpson, "National Park Logs First Grizzly Attack," *Jackson Hole Guide*, August 17, 1994, 7; Diana Hingston, "Park City Man Mauled by Grizzly Bear in the Grand Teton Park," *Park* [City UT] *Record*, August 18, 1994, 1; and Gunther et al., "Grizzly Bear–Human Conflicts 1994–1997," 14.

16. Christopher Shelton, "Grizzly Bear Mauls Hunter Near Moran," *Jackson Hole Guide*, September 21, 1994, 1; Angus M. Thuermer Jr., "Bear Bites Californian Hiking in Yellowstone," *Jackson Hole News*, September 28, 1994, 8; and Gunther et al., "Grizzly Bear–Human Conflicts 1994–1997," 14.

17. "Case Incident Record 91843," July 24, 1997, GTNP, Ursids, Series 10.4, Box 23, File Unit 10.

18. "Case Incident Record" and "Supplementary Case Incident Record 971719," July 20, 1997, including statements of victim and river guide, and park press releases, GTNP, Ursids, Series 10.4, Box 23, File Unit 2.

19. "Case Incident Record" and "Supplementary Case Incident Record 973008," September 1, 1997, including chronology of events and a Teton County EMS report, GTNP, Ursids, Series 10.4, Box 23, File Unit 10.

20. "Case Incident Record" and "Supplementary Case Incident Record 980841," June 5, 1998, including handwritten statement of witness and photographs, GTNP, Ursids, Series 10.4, Box 23, File Unit 10.

21. Grand Teton National Park, *Black Bear/Human Conflicts Resulting in Human Injury*, no author or date, GTNP, Ursids, Series 10, File Unit 16.

22. David Simpson, "Man Mauled by Grizzly in Teton Park," *Jackson Hole Guide*, September 2, 1998, 1, A23.

23. Schullery, *Yellowstone's Ski Pioneers*, 139–42.

24. Gunther et al., "Grizzly Bear-Human Conflicts, Confrontations, and Management Actions in the Yellowstone Ecosystem, 2001," 83–84.

25. "Case Incident Record" and "Supplementary Case Incident Record 010348," April 14, 2001, GTNP, Ursids, Series 10.4, Box 23, File Unit 12.

26. Gunther et al., "Grizzly Bear–Human Conflicts, Confrontations, and Management Actions in the Yellowstone Ecosystem, 2001," 83–84.

27. Cain, "2007 Wildlife Management: Human–Bear Conflicts." Also, Cory Hatch, "Grizz Attack 'a Warning,'" *Jackson Hole News and Guide*, June 20, 2007, 1, 27A.

28. Cain, "2008 Wildlife Management: Human–Bear Conflicts."

29. Stephenson and Cain, "Wildlife Management: Human-Bear Interface." Also, Cory Hatch, "Attack Stokes Criticism of Hunt," *Jackson Hole News and Guide*, November 3, 2011, 1, 25A.

30. Billy Arnold, "Grizzly in Attack Won't Be Punished," *Jackson Hole News and Guide*, May 22, 2024, 1, 14A.

13. GUNS, SNARES, AND BEARS

1. Email from Larry Rockefeller (with embedded message from Mark Rockefeller) to author, February 19, 2021. Used with permission.

2. Crockett, "The Prehistoric Peoples of Jackson Hole," 23–37.

3. "Fighting to Save Our Game," *Philadelphia Inquirer*, December 23, 1917, 41.

4. Blair, *The History of Wildlife Management in Wyoming*, 58.

5. Diem, Diem, and Lawrence, *Tale of Dough Gods, Bear Grease, Cantaloupe, and Sucker Oil*.

6. Emily I. Anderson, "Oral Leek's Memories of Life in Jackson Hole," *Jackson Hole Guide*, May 17, 1973, 17.

7. Diem, Diem, and Lawrence, *Tale of Dough Gods, Bear Grease, Cantaloupe, and Sucker Oil*, 43, 80.

8. "Wild Life of Wyoming Fed in Winter to Perpetuate Species; Thousands Fed," *Jackson's Hole Courier*, October 19, 1929, 4.

9. "Game Licenses Yield $100,000," *Jackson's Hole Courier*, December 5, 1929, 1.

10. "Wyoming Mecca for Big Game Hunters," *The Minden* [Nebraska] *Courier*, October 30, 1930, 9.

11. "Game Sensus-Made on Forests," *Jackson's Hole Courier*, March 19, 1931, 1.

12. "Such Is Fame," *Jackson's Hole Courier*, June 1, 1933, 1; and "Clark Gable, Star, Got 'Limit' Every Day While Fishing," *Jackson's Hole Courier*, June 8, 1933, 1.

13. "Kelly" news column, *Jackson's Hole Courier*, May 23, 1940, 3.

14. "Sport (?)," *Jackson's Hole Courier*, May 28, 1931, 4. (Attributed to *Wind River Mountaineer*, copies from 1932 not available.)

15. Blair, *The History of Wildlife Management in Wyoming*, 89, 118.

16. Photo 2005.0163.004, JHHSM.

17. "Big Grizzly Taken," *Jackson Hole Guide*, October 13, 1960, 16; and "Nymeyer's Bear Is Prize Trophy," *Jackson Hole Guide*, May 15, 1961, 5.

18. Reneau and Buckner, *Records of North American Big Game*, 12th ed., 172–73.

19. "History of Park Extension Efforts Shows That Field Committees Opposed," *Jackson's Hole Courier*, January 20, 1938, 1.

20. "Bagley Tells Wardens to Recognize Only State Laws," *Jackson's Hole Courier*, May 13, 1943, 1.

21. Paul R. Franke, *Annual Wildlife Report*, October 3, 1944, NARA-KC, Record Group 79, Box 77, File 207.01.

22. Blair, *The History of Wildlife Management in Wyoming*, 41.

23. Diem, Diem, and Lawrence, *Tale of Dough Gods, Bear Grease, Cantaloupe, and Sucker Oil*, 79–80.

24. "Bear Pays with Life for Stealing Quarter of Beef," *Jackson's Hole Courier*, October 17, 1935, 1.

25. "Big Game Population Estimated at 200,000," *Jackson's Hole Courier*, April 20, 1950, 9.

26. "Wildlife Populations Said Steady in National Parks," *Jackson's Hole Courier*, March 9, 1950, 10.

27. "First Bear Killed," *Jackson's Hole Courier*, May 22, 1952, 1.

28. "Big Grizzly Shot," *Jackson Hole Guide*, September 18, 1952, 1.

29. "G-F Commission Considers Protection of Grizzly," *Jackson's Hole Courier*, August 20, 1959, 3.

30. Craighead et al., *Grizzly Bear Mortalities in the Yellowstone Ecosystem 1959–1987*, Appendix.
31. Blair, *The History of Wildlife Management in Wyoming*, 226.
32. National Research Council, *Report of the Committee on Yellowstone Grizzlies*, 39–42.
33. J. Philip Magers, "State Says Bears Aren't Endangered," *Billings Gazette*, April 13, 1974, 8; and "Wyoming Favors Grizzly Moratorium," *The Billings Gazette*, May 11, 1974, 15.

14. THE (NOT-SO-) FORGOTTEN PARKWAY

1. Steely, *Administrative History*, 67, 73.
2. U.S. Department of the Interior, National Park Service, *Management Policies*, 2006, 98–101.
3. *An Act to Establish a National Park Service*.
4. Wyoming Game and Fish Department, "Black Bear Harvest Report."
5. Craighead et al., *Grizzly Bear Mortalities*.
6. U.S. Department of the Interior, Fish and Wildlife Service, *Proposed Revision of Special Regulations for the Grizzly Bear*, 42525.
7. Carole Legg, "Proposal Calls for Nuisance Grizzly Hunts," *Casper Star-Tribune*, October 6, 1989, 11.
8. Angus Thuermer Jr., "Nuisance Griz Hunt May Be Headed for Federal Approval," *Jackson Hole News*, December 28, 1989, 3.
9. "Unwarranted," letter by Wyoming Game and Fish Department director Francis E. "Pete" Petera, *Jackson Hole Guide*, April 11, 1990, 5.
10. Sherry Devlin, "The Wily Bureaucrat: Earth First! to Protest Bear Hunt," *The Missoulian*, February 20, 1990, 7.
11. Richard Abendroth, "Grizzly Plan Creates Stir," *Jackson Hole Guide*, November 24, 1993, 1, 14.
12. U.S. Department of the Interior, National Park Service, *Environmental Assessment and Finding of No Significant Impact, Flagg Ranch Development Plan*.
13. U.S. Department of the Interior, National Park Service, *Environmental Assessment and Finding of No Significant Impact, Flagg Ranch Development Plan*, Appendix A, 79–83.
14. U.S. Fish and Wildlife Service: https://www.fws.gov/species/grizzly-bear-ursus-arctos-horribilis, and Montana Public Radio: https://www.mtpr.org/montana-news/2021ñ04ñ02/timeline-a-history-of-grizzly-bear-recovery-in-the-lower-48-states.
15. U.S. Fish and Wildlife Service Environmental Conservation Online System, March 2024, https://ecos.fws.gov/ecp/boxscore.
16. U.S. Fish and Wildlife Service, *Grizzly Bear Recovery Program 2022 Annual Report*, 3, https://www.fws.gov/sites/default/files/documents/2022%20GBRP%20Annual%20Report.pdf.

17. Masica, intermountain regional director, National Park Service, to director, U.S. Fish and Wildlife Service, "NPS Comments on Proposed Rule, Removing the GYE Population of Grizzly Bears from the Federal List of Endangered and Threatened Wildlife and Final Draft Conservation Strategy." Docket is fws-r6-es-2016–0042, May 10, 2016, on file at Grand Teton National Park.

18. Kurt Repanshek, "Wyoming Has No 'Intent' to Allow Grizzly Hunt on NPS Lands If Grizzlies Delisted," *National Parks Traveler*, November 29, 2016, 3.

19. U.S. Fish and Wildlife Service, https://www.fws.gov/species/grizzly-bear-ursus -arctos-horribilis.

20. Brian Miller, "Gianforte, FWP Say Grizzly Translocations Shows [*sic*] Montana Ready for Delisting," *Daily Montanan*, August 5, 2024.

15. SCIENCE AND TETON BEARS

1. Conversation with Dr. Henry J. Harlow, March 25, 2022, Tucson AZ. Used with permission.

2. Craighead, Sumner, and Mitchell, *The Grizzly Bears of Yellowstone*, 105–6.

3. Berger, Swenson, and Persson, "Recolonizing Carnivores and Naïve Prey," 1036–39.

4. Berger, "Fear, Human Shields and the Redistribution of Prey and Predators," 622–23.

5. Pyare and Berger, "Beyond Demography and Delisting," 67–71.

6. Holm, Lindzey, and Moody, "Interactions of Sympatric Black and Grizzly Bears in Northwest Wyoming," 105–6.

7. Costello et. al., "Diet and Macronutrient Optimization in Wild Ursids," 16–18.

8. Frattaroli, "Black Bear Ecology in Southern Grand Teton National Park," 27–29, 131.

9. Costello et al., *Impacts of a Multi-use Pathway on American Black Bears*, 23–25.

10. van Manen et al., "Primarily Resident Grizzly Bears Respond to Late-Season Elk Harvest," 8, 10–12.

11. Nelson et al., "An Evaluation of the Be Bear Aware Program," 71–78.

12. Blotkamp, "Day Hikers in Bear Country," 89–102, 121–22.

16. BEARS IN THE 'HOOD

1. Conversation with Barry Alexander, retired Grand Teton National Park heavy equipment operator, August 23, 2021. Used with permission.

2. Schwartz and Haroldson, "Hazards Affecting Grizzly Bear Survival," 661.

3. Mattson, Knight, and Blanchard, "The Effects of Developments and Primary Roads on Grizzly Bear Habitat Use," 262–63.

4. Mealey, *Interagency Grizzly Bear Guidelines*, 4.

5. U.S. Department of the Interior, National Park Service, *Master Plan for Grand Teton National Park*.

6. Herrero, *Bear Attacks*, 41.

7. "Official Proceedings of the Board of County Commissioners, Teton County, Wyoming," *Jackson Hole Guide*, September 19, 1985, 26.

8. Kersten Swinyard, "Veteran Commissioners Hit Home Stretch," *Jackson Hole News and Guide*, December 24, 2002, 21.

9. Gunther et al., "Managing Human-Habituated Bears to Enhance Survival," 374.

10. Stoen et al., "Physiological Evidence for a Human-Induced Landscape of Fear in Brown Bears," 246–47.

11. Gunther et al., "Habituated Grizzly Bears," 36.

12. Jim Stanford, "Rival Plans Touted for Park Paths," *Jackson Hole News and Guide*, July 13, 2005, 1, 42.

13. U.S. Department of the Interior, National Park Service, *Transportation Plan*, 3–6.

14. Cory Hatch, "Old Grizzly Put Down; Had Gone Near Homes," *Jackson Hole News and Guide*, May 27, 2009, 22.

15. Kidd and Monz, "Understanding and Managing Wildlife Jams," 75–76, 78–79.

16. U.S. Department of the Interior, National Park Service, *Record of Decision (ROD) for the Moose-Wilson Corridor*, 2–6.

17. Gunther et al., "Habituated Grizzly Bears: A Natural Response to Increasing Visitation in Yellowstone and Grand Teton National Parks," 36.

18. Personal communication and copy shared of IGBST biologist Mark A. Haroldson's 2008 talk on bears, presented in Grand Teton National Park.

19. Herrero et al., "From the Field: Brown Bear Habituation to People," 366–67.

20. Richardson et al., "The Economics of Roadside Bear-Viewing," 105–10.

21. Gunther et al., "Managing Human-Habituated Bears to Enhance Survival," 378.

22. Richardson et al., "Visitor Perceptions of Roadside Bear Viewing," 304.

23. Herrero et al., "From the Field: Brown Bear Habituation to People," 365–69.

17. MAKING A DIFFERENCE

1. Jackson Hole Wildlife Foundation, "2023 Community Impact Report," https://jhwildlife.org/wp-content/uploads/2024/03/2023-Annual-jhwf-Community-Impact-Report.pdf.

2. Conversation with Mallory Smith, chief of business resources, Grand Teton National Park, August 28, 2022.

3. Conversation with Kate Wilmot, branch chief for Fish and Wildlife Department, Grand Teton National Park, January 5, 2023, and subsequent email on May 23, 2023.

4. Conversation with Jeff Willemain, July 24, 2022. Used with permission.

5. Conversation with Steve Kilpatrick, retired Wyoming Game and Fish Department biologist, July 7, 2022. Used with permission.

6. Turbak, "Swapping Conflict for Conservation," 2004.

7. National Wildlife Federation, "The National Wildlife Federation's Grazing Agreements," 2016.

8. Fence removal project records 1996–2007, provided by Kate Gersh, Jackson Hole Wildlife Foundation.

9. Conversation with Chuck Schneebeck, February 7, 2023. Used with permission.

18. CELEBRITY

1. Joe Szuszwalak, "Jackson Area Residents and Visitor Actions Are Vital to Survival of Bears—Including 399 and Offspring," U.S. Fish and Wildlife Service press release, April 6, 2022.

2. Conversation with Michelle Weber, Jackson, Wyoming, police chief, July 28, 2022, Jackson WY. Shared with permission. Also, "Caught on Video: Mama Bear 399 and Cubs Go to Jail," *Jackson Hole News and Guide* online video post, November 10, 2021.

3. "Grizzly Bear Halts Traffic on Teton Pass Saturday," *Jackson's Hole Courier*, August 28, 1930, 1.

4. Concerned Citizens for the Elk, Ad: "TOO MANY GRIZZLIES," *Jackson Hole News and Guide,* June 13, 2012, 27.

5. Haroldson et al., "Grizzly Cub Adoptions Confirmed in Yellowstone and Grand Teton National Parks," 58, 61.

6. Presentation by Kate Wilmot, Grand Teton National Park bear management specialist, to the Grand Teton National Park Foundation and other listeners via video link, April 15, 2021, available online at gtnpf.org; and conversation with Kate Wilmot on January 5, 2023; also U.S. Geological Survey, Interagency Grizzly Bear Study Team, *Annual Reports 2001–2021.*

7. Conversation with Frank van Manen and Mark Haroldson, IGBST research biologists, June 1, 2022.

8. Conversation with Frank van Manen and Mark Haroldson, June 1, 2022.

9. Susan M. Kahn, "Different Category" letter to *Jackson Hole News and Guide*, November 16, 2011, 4.

10. Ryan Dorgan, "Grizzly Managers Worried about 399 and Cubs," *Jackson Hole Daily*, April 11, 2022.

11. Angus M. Thuermer Jr. "Officials Kill Sub-adult Offspring of Famous Grizzly 399," *WyoFile*, July 14, 2022.

12. Mark Heinz, "Grizzly 399 Lives! And She's Got a New Cub!" *Cowboy State Daily*, May 16, 2023.

13. Sadowski, *Bearing the Burden*, 14, 18–21.

14. Mike Koshmrl, "Homeowner Feeds Teton Park Grizzlies for Years; Feds Decline Charges," *Jackson Hole News and Guide*, February 17, 2021.

15. Eric Reinertson, "Antisocial Media," 34.

16. Mike Koshmrl, "Park: 399's Cub Dead," *Jackson Hole News and Guide*, June 20, 2016.

17. Katie Hill, "Californians Are Losing It After the Death of the 'King' Puma. That's Not a Good Thing for Mountain Lion Conservation," *Outdoor Life* online, December 23, 2022.
18. Mattson, Logan, and Sweanor, "Factors Governing Risk of Cougar Attacks on Humans," 135.

19. THE UNKNOWABLE BEARS

1. Knight, Blanchard, and Haroldson, *Yellowstone Grizzly Bear Investigation*, 6–7.
2. Brasington, *A Yellowstone Grizzly Trilogy*.
3. Haroldson et al., *What to Do with Offspring of Conflict Bears*.
4. Meagher and Fowler, "The Consequences of Protecting Problem Grizzly Bears," 133–34.
5. Schwartz and Haroldson, *Yellowstone Grizzly Bear Investigations*, 9; Schwartz, Haroldson, and West, *Yellowstone Grizzly Bear Investigations 2007*, 7; and *2008*, 7; and van Manen, et al., *Yellowstone Grizzly Bear Investigations 2013*, 10. Also, Jeff Obrecht, "The Life and Times of Grizzly Bear 179," *Wyomingnews.com*, posted online March 10, 2014.

Bibliography

ARCHIVES

GTNP. Grand Teton National Park Archives, National Park Service, Moose WY.

IGBC-YES. Records of the Interagency Grizzly Bear Committee, Yellowstone Ecosystem Managers' Subcommittee. Gallatin National Forest headquarters, Bozeman MT.

JHHSM. Jackson Hole Historical Society and Museum, Jackson WY.

NARA. National Archives and Records Administration, Washington DC.

NARA-KC. National Archives and Records Administration, Record Group 79, Records of the National Park Service, Kansas City KS.

NPS. National Park Service Oral History Collection (HFCA 1817), Harpers Ferry Center WV.

BOOKS AND ARTICLES

Albright, Horace M., and Robert Cahn. *Birth of the National Park Service*. Salt Lake City: Howe Brothers, 1985.

An Act to Establish a National Park Service, and for Other Purposes (National Park Service Organic Act). Public Law 64–235, *U.S. Statutes at Large*, Vol. 39, Part 1, Chap. 408 (August 25, 1916): 535–36.

An Act to Establish a New Grand Teton National Park in the State of Wyoming, and for Other Purposes. Public Law 81–787, *U.S. Statutes at Large* 64 (1950): 849–53.

Allen, Durward L., Larry Erickson, E. Raymond Hall, and Walter M. Schirra. "Report to Secretary of the Interior James G. Watt: A Review and Recommendations on Animal Problems and Related Management Needs in Units of the National Park System. (A Report from the Special Task Force of the National Park System Advisory Board and Its Council)," October 7, 1981. *The George Wright Forum* 1, no. 1 (Autumn 1981): 9–33.

Anderson, Charles R., Mark A. Ternant, and David S. Moody. "Grizzly Bear–Cattle Interactions on Two Grazing Allotments in Northwest Wyoming." *Ursus* 13 (2002): 247–56.

Baillie-Grohman, William A. *Camps in the Rockies. Being a Narrative of Life on the Frontier, and Sport in the Rocky Mountains, with an Account of the Cattle Ranches of the West*. New York: Charles Scribner's Sons, 1882.

Barker, Rocky. *Saving All the Parts: Reconciling Economics and the Endangered Species Act*. Washington DC: Island Press, 1993.

Berger, Joel. *The Better to Eat You With: Fear in the Animal World*. Chicago: University of Chicago Press, 2008.

Berger, Joel. "Fear, Human Shields, and the Redistribution of Prey and Predators in Protected Areas." *Biology Letters* 3 (2007): 620–23.

Berger, Joel, Jon E. Swenson, and Inga-Lill Persson. "Recolonizing Carnivores and Naïve Prey: Conservation Lessons from Pleistocene Extinctions." *Science* 291 (2001): 1036–39.

Blair, Neal. *The History of Wildlife Management in Wyoming*. Cheyenne: Wyoming Game and Fish Department, 1987.

Blanchard, Bonnie M., Richard R. Knight, and E. M. Yount. "Mortality." *Yellowstone Grizzly Bear Investigations: Annual Report of the Interagency Grizzly Bear Study Team, 1978–1979* (1980): 20–23.

Blotkamp, Ariel J. "Day Hikers in Bear Country: A Study of Knowledge, Fear, and Protection Motivation." MS thesis, University of Idaho, Moscow, 2011.

Bonney, Orrin H., and Lorraine G. Bonney. *The Grand Controversy: The Pioneering Climbs in the Teton Range and the Controversial First Ascent of the Grand Teton*, New York: American Alpine Club Press, 1992.

Brasington, Tyler. *A Yellowstone Grizzly Trilogy: Three Grizzly Life Histories Linked by Lineage*. University of Wisconsin, Stevens Point, December 15, 2020. www.yellowstonegrizzlyproject.org.

Cain, Steven L., ed. "2007 Wildlife Management: Human-Bear Conflicts." In *2007 Wildlife Conservation, Management, and Research*. On file at the Division of Science and Resource Management, Grand Teton National Park, Moose WY.

Cain, Steven L., ed. "2008 Wildlife Management: Human-Bear Conflicts." In *2008 Wildlife Conservation, Management, and Research*. On file at the Division of Science and Resource Management, Grand Teton National Park, Moose WY.

Consolo-Murphy, Sue, ed. "Holding on to Yellowstone's Grizzlies: A Parting Chat with a 24-Year Veteran of Yellowstone's Grizzly Bear Wars." *Yellowstone Science* 6, no. 1 (Winter 1998): 2–9.

Costello, Cecily M., Steven L. Cain, Ryan M. Neilson, Christopher Servheen, and Charles C. Schwartz. *Impacts of a Multi-use Pathway on American Black Bears in Grand Teton National Park*, WY. 2013. On file at Grand Teton National Park WY and available online at umt.edu.

Costello, Cecily M., Steven L. Cain, Shannon Pils, Leslie Frattaroli, Mark A. Haroldson, and Frank T. van Manen. "Diet and Macronutrient Optimization in Wild Ursids: A Comparison of Grizzly Bears with Sympatric and Allopatric Black Bears." *PLosOne* 11, no. 5 (May 18, 2016): e0153702. https://doi.org/10.1371/journal.pone.0153702.

Craighead, John J. "A Proposed Delineation of Critical Grizzly Bear Habitat in the

Yellowstone Region." Monograph presented at the Fourth International Conference on Bear Research and Management, Kalispell MT, February 1977. Bear Biology Association Monograph Series, no. 1, 1980.

Craighead, John J., Kenneth R. Greer, Richard R. Knight, and Helga I. Pac. *Grizzly Bear Mortalities in the Yellowstone Ecosystem 1959–1987* (November 1988). Helena: Montana Department of Fish, Wildlife and Parks (also the Craighead Wildlife-Wildlands Institute, Interagency Grizzly Bear Study Team, and National Fish and Wildlife Foundation). On file with the U.S. Geological Survey, Interagency Grizzly Bear Study Team, Bozeman MT.

Craighead, John J., Jay S. Sumner, and John A. Mitchell. *The Grizzly Bears of Yellowstone: Their Ecology in the Yellowstone Ecosystem, 1959–1992.* Washington DC: Island Press, 1995.

Crockett, Stephanie. "The Prehistoric Peoples of Jackson Hole." In *A Place Called Jackson Hole: The Historic Resource Study of Grand Teton National Park*, edited by John Daugherty, 21–41. Moose WY: National Park Service, Grand Teton National Park, 1999.

Daugherty, John, with contributions by Stephanie Crockett, William H. Goetzmann, and Reynold G. Jackson. *A Place Called Jackson Hole: The Historic Resource Study of Grand Teton National Park.* Moose WY: National Park Service, Grand Teton National Park, 1999.

Diem, Kenneth L., Lenore L. Diem, and William C. Lawrence. *A Tale of Dough Gods, Bear Grease, Cantaloupe, and Sucker Oil: Marymere/Pinetree/Mae-Lou/AMK Ranch.* Laramie: University of Wyoming–National Park Service Research Center, 1986.

Fisher Smith, Jordan. *Engineering Eden: A Violent Death, a Federal Trial, and the Struggle to Restore Nature in Our National Parks.* New York: Crown Publishing: 2016.

Frattaroli, Leslie M. "Black Bear Ecology in Southern Grand Teton National Park." MS thesis, University of Montana, Missoula, 2011.

Grand Teton National Park. *Biological Assessment, Threatened and Endangered Species for Reissuance of Grazing Permits in Grand Teton National Park.* March 2005. Moose WY.

Grand Teton National Park and John D. Rockefeller, Jr. Memorial Parkway. *Natural and Cultural Resources Vital Signs 2013–2021.* Moose WY.

Gunther, Kerry A., Mark Bruscino, Steve Cain, Ted Chu, Kevin Frey, and Richard R. Knight. "Grizzly Bear–Human Conflicts, Confrontations, and Management Actions in the Yellowstone Ecosystem, 1994–1996." Interagency Grizzly Bear Committee, Yellowstone Ecosystem Subcommittee Reports. Yellowstone National Park WY, 1995–1997.

Gunther, Kerry A., Mark Bruscino, Steve Cain, Jeff Copeland, Kevin Frey, Mark A. Haroldson, and Charles C. Schwartz. "Grizzly Bear–Human Conflicts, Confrontations, and Management Actions in the Yellowstone Ecosystem, 2000." In *Yellowstone Grizzly Bear Investigations: Annual Report of the Interagency Grizzly Bear Study*

Team, 2000, edited by Charles C. Schwartz and Mark A. Haroldson, 64–107. Bozeman MT: U.S. Geological Survey, 2001.

Gunther, Kerry A., Mark Bruscino, Steve Cain, Lauri Hanauska-Brown, Kevin Frey, Mark A. Haroldson, and Charles C. Schwartz. "Grizzly Bear–Human Conflicts, Confrontations, and Management Actions in the Yellowstone Ecosystem, 2001." In *Yellowstone Grizzly Bear Investigations: Annual Report of the Interagency Grizzly Bear Study Team, 2001*, edited by Charles C. Schwartz and Mark A. Haroldson, 56–93. Bozeman MT: U.S. Geological Survey, 2002.

Gunther, Kerry A., Rebecca R. Shoemaker, Kevin L. Frey, Mark A. Haroldson, Steven L. Cain, Frank T. van Manen, and Jennifer K. Fortin. "Dietary Breadth of Grizzly Bears in the Greater Yellowstone Ecosystem." *Ursus* 25, no. 1 (2014): 61–73.

Gunther, Kerry A., Katharine R. Wilmot, Steven L. Cain, Travis Wyman, Eric G. Reinertson, and Amanda M. Bramblett. "Habituated Grizzly Bears: A Natural Response to Increasing Visitation in Yellowstone and Grand Teton National Parks." *Yellowstone Science* 23, no. 2 (2015): 33–40.

Gunther, Kerry A., Katharine R. Wilmot, Steven L. Cain, Travis Wyman, Eric G. Reinertson, and Amanda M. Bramblett. "Managing Human-Habituated Bears to Enhance Survival, Habitat Effectiveness, and Public Viewing." *Human-Wildlife Interactions* 12, no. 3 (2018): 373–86.

Hanks, Allyn F. "Talus." *Grand Teton Nature Notes* 3, no. 3 (1937): 35. https://archive.org/details/naturenotesgrand28unit/page/n53/mode/2up.

Haroldson, Mark A., Kerry A. Gunther, Daniel L. Bjornlie, Daniel J. Thompson, Kevin L. Frey, and Bryan C. Aber. *What to Do with Offspring of Conflict Bears: Genetic Insights from the Greater Yellowstone Ecosystem*. Presentation to the semi-annual Yellowstone Ecosystem Grizzly Bear Managers' Subcommittee, October 29, 2014, Bozeman MT.

Haroldson, Mark A., Kerry A. Gunther, Steven L. Cain, Katharine R. Wilmot, and Travis Wyman. "Grizzly Cub Adoption Confirmed in Yellowstone and Grand Teton National Parks." *Yellowstone Science* 23, no. 2 (2015): 58–61.

Haroldson, Mark A., Mark A. Ternent, Kerry A. Gunther, and Charles C. Schwartz. "Grizzly Bear Denning Chronology and Movements in the Greater Yellowstone Ecosystem." *Ursus* 13 (2002): 29–37.

Haroldson, Mark A., Mark Ternent, Greg Holm, Roger A. Swalley, Shannon Podruzny, David Moody, and Charles C. Schwartz. *Yellowstone Grizzly Bear Investigations: Annual Report of the Interagency Grizzly Bear Study Team, 1997*. Bozeman MT: U.S. Geological Survey, 1998.

Herrero, Stephen. *Bear Attacks: Their Causes and Avoidance*. Guilford CT: Lyons Press, 2002.

Herrero, Stephen, Andrew Higgins, James E. Cordoza, Laura I. Hajduk, and Tom S. Smith. "Fatal Attacks by American Black Bear on People: 1900–2009." *Journal of Wildlife Management* 75, no. 3 (2011): 596–603.

Herrero, Stephen, Tom Smith, Terry D. DeBruyn, Kerry Gunther, and Colleen A. Matt. "From the Field: Brown Bear Habituation to People—Safety, Risks, and Benefits." *Wildlife Society Bulletin* 33, no. 1 (2005): 362–73.

Hoak, John H., Tim W. Clark, and Bob Wood. "Grizzly Bear Distribution, Grand Teton National Park Area, Wyoming." *Northwest Science* 55, no. 4 (1981): 245–47.

Holm, Greg W., Frederick G. Lindzey, and David S. Moody. "Interactions of Sympatric Black and Grizzly Bears in Northwest Wyoming." *Ursus* 11 (1999): 99–108.

Humstone, Mary. "Elk Ranch Determination of Eligibility for the National Register of Historic Places." *University of Wyoming National Park Service Research Center Annual Report* 33, article 5 (2010): 37–46.

Huyler, Jack. *And That's the Way It Was in Jackson's Hole*. 2nd ed. Jackson WY: Jackson Hole Historical Society and Museum, 2003.

Interagency Grizzly Bear Study Team. *Bear-Caused Human Fatalities in the Greater Yellowstone Ecosystem, 1892–Present*, June 7, 2023.

Johnson, Lady Bird. Daily Diary Entry, September 9, 1965. Lady Bird Johnson's White House Diary Collection, LBJ Presidential Library, National Archives and Records Administration, Austin TX. https://www.discoverlbj.org/item/ctjd-19650909.

Johnson, S. J., and D. E. Griffel. "Sheep Losses on Grizzly Bear Range." *Journal of Wildlife Management* 46, no. 3 (1982): 786–90.

Jorgensen, Carole J. "Bear-Livestock Interactions, Targhee National Forest." MS thesis, University of Montana, Missoula, 1979.

Jorgensen, Carole J. "Bear-Sheep Interactions, Targhee National Forest." In *Bears: Their Biology and Management: Fifth International Conference on Bear Research and Management: A Selection of Papers from the Conference Held at Madison, Wisconsin, USA, February 1980*, edited by E. Charles Meslow, 191–200. Bozeman MT: International Association for Bear Research and Management, 1983.

Judge, Francis. "Vital Laughter." *Atlantic Monthly* (July 1954): 47–52.

Kidd, Abigail M., and Christopher Monz. "Understanding and Managing Wildlife Jams in National Parks: An Evaluation in Grand Teton National Park." *University of Wyoming— National Park Service Research Station Annual Report* vol. 39 (2016): 73–80.

Knight, Richard R., Bonnie M. Blanchard, and Mark A. Haroldson. *Yellowstone Grizzly Bear Investigations: Annual Report of the Interagency Grizzly Bear Study Team, 1996*. Bozeman MT: U.S. Geological Survey, 1997.

Knight, Richard R., Bonnie M. Blanchard, and Katherine C. Kendall. *Yellowstone Grizzly Bear Investigations: Annual Report of the Interagency Grizzly Bear Study Team, 1980*. Bozeman MT: U.S. Geological Survey, 1981.

Knight, Richard R., Bonnie M. Blanchard, Katherine C. Kendall, and Lloyd E. Oldenburg. *Yellowstone Grizzly Bear Investigations: Annual Report of the Interagency Grizzly Bear Study Team, 1978–79*. Bozeman MT: U.S. Geological Survey, 1980.

Knight, Richard R., and Stephen L. Judd. "Grizzly Bears That Kill Livestock." In *Bears: Their Biology and Management*. Vol. 5, *A Selection of Papers from the Fifth*

International Conference on Bear Research and Management, Madison, Wisconsin, USA, February 1980, 186–190. West Glacier MT: International Association for Bear Research and Management, 1983.

Knight, Richard R., and Stephen L. Judd. "Grizzly Bears That Kill Livestock." In *Yellowstone Grizzly Bear Investigations: Annual Report of the Grizzly Bear Study Team, 1978–1979*, 43–50. Bozeman MT: U.S. Geological Survey, 1980.

Mangelsen, Thomas D., and Todd Wilkinson. *Grizzly 399: The World's Most Famous Mother Bear*. New York: Rizzoli International Publications, 2023.

Mangelsen, Thomas D., and Todd Wilkinson. *Grizzly, The Bears of Greater Yellowstone: Grizzly 399 and Her Family of Pilgrim Creek*. New York: Rizzoli International Publications, 2015.

Mattson, David D., Richard R. Knight, and Bonnie M. Blanchard. "The Effects of Developments and Primary Roads on Grizzly Bear Habitat Use in Yellowstone National Park, Wyoming." *International Conference on Bear Research and Management* 7 (1987): 259–73.

Mattson, David, Kenneth Logan, and Linda Sweanor. "Factors Governing Risk of Cougar Attacks on Humans." *Human-Wildlife Interactions* 5, issue 1, article 15 (Spring 2011):135–58. https://doi.org/10.26077/sey6-hq10.

Meagher, Mary, and Sandy J. Fowler. "The Consequences of Protecting Problem Grizzly Bears." *Bear-People Conflicts, Proceedings of a (1986) Symposium on Management Strategies*, Yellowknife, Northwest Territories, Canada (1989): 141–44.

Mealey, Stephen P., ed. *Interagency Grizzly Bear Guidelines*. U.S. Department of Agriculture, Forest Service; U.S. Department of Interior, Bureau of Land Management, U.S. Fish and Wildlife Service, National Park Service; Idaho Department of Fish and Game; Montana Department of Fish, Wildlife and Parks; Washington Game Department; and the Wyoming Game and Fish Department, 1986. http://npshistory.com/publications/wildlife/interagency-grizzly-bear-guidelines.pdf.

Miller, Brian, and Amanda Hardy. *Final Project Report for "Integrating Climate Considerations into Grazing Management Programs in National Parks."* March 29, 2022. U.S. Department of Interior, National Park Service, Climate Change Response Program. https://www.sciencebase.gov/catalog/item/5cf6fba8e4b0d63728b9b4cc.

Muir, John. "Among the Animals of the Yosemite." *Atlantic Monthly* 82, no. 493 (November 1898): 617–31.

Murie, Adolph. "Cattle on Grizzly Bear Range." *Journal of Wildlife Management* 12, no. 1, (1948): 57–72.

National Research Council. *Report of the Committee on Yellowstone Grizzlies*. Washington DC: National Academies Press, 1974. https://doi.org/10.17226/20128.

National Wildlife Federation. "The National Wildlife Federation's Grazing Agreements," December 21, 2016. https://www.nwf.org/~/media/PDFs/Regional/Northern-Rockies/Wildlife-Conflict-Resolution-Program/12-21-16_WCR-Grazing-Allotment-Map.ashx.

Nelson, Nanette M., Patricia A. Taylor, Tyler Hopkins, and Amy Rieser. "An Evaluation of the Be Bear Aware Program at Grand Teton National Park." *University of
Wyoming National Park Service Research Center Annual Report* 33, article 9 (2010):
67–80. Grand Teton National Park, Moose WY; and the University of Wyoming,
Laramie.

O'Ney, Susan, and Rob Gipson. "A Century of Fisheries Management in Grand Teton
National Park." In *Proceedings, Greater Yellowstone Science Conference*, 131–134. 2005.

Pelton, Michael R. "Black Bear." In *Wild Mammals of North America: Biology, Management, and Conservation*, edited by George A. Feldhamer, Bruce C. Thompson, and
Joseph A. Chapman, 547–55. Baltimore: Johns Hopkins University Press, 2003.

Pyare, Sanjay, and Joel Berger. "Beyond Demography and Delisting: Ecological Recovery for Yellowstone's Grizzly Bears and Wolves." *Biological Conservation* 113 (2003):
63–73.

Pyare, Sanjay, Steve Cain, Dave Moody, Chuck Schwartz, and Joel Berger. "Carnivore Re-
colonisation: Reality, Possibility and a Non-equilibrium Century for Grizzly Bears in
the Southern Yellowstone Ecosystem." *Animal Conservation* 7 (2004): 71–77.

Reinertson, Eric. "Antisocial Media." In *Dangerous Beauty: Encounters with Grizzlies
and Bison in Yellowstone*, edited by Sandy Sisti and Carolyn Jourdan, 34. Chicago:
Zo'o Media, 2017.

Reneau, Jack, and Eldon L. "Buck" Buckner, ed. *Records of North American Big Game*,
12th edition. Missoula MT: Boone and Crockett Club, 2005.

Richardson, Leslie, Kerry Gunther, Tatjana Rosen, and Chuck Schwartz. "Visitor Perceptions of Roadside Bear Viewing and Management in Yellowstone National Park."
The George Wright Forum 32, no. 3 (2015): 299–307.

Richardson, Leslie, Tatjana Rosen, Kerry Gunther, and Chuck Schwartz. "The Economics of Roadside Bear-Viewing." *Journal of Environmental Management* 140 (2014):
102–110.

Righter, Robert W. *Crucible for Conservation: The Struggle for Grand Teton National
Park*. Moose WY: Grand Teton Association, 1982.

Righter, Robert W. *Peaks, Politics and Passion: Grand Teton National Park Comes of Age*.
Moose WY: Grand Teton Association, 2014.

Robbins, Charles T., Merav Ben-David, Jennifer K. Fortin, and O. Lynne Nelson.
"Maternal Condition Determines Birth Date and Condition of Newborn Bear
Cubs." *Journal of Mammalogy* 93, no. 2 (2012): 540–546.

Rockwell, David. *Giving Voice to Bear*. Niwot CO: Roberts Rinehart Publishers, 1991.

Russell, Andy. *Grizzly Country*. New York: Alfred A. Knopf, 1967.

Sadowski, Lauren E. *Bearing the Burden: Multidisciplinary Integration in Human-Bear
Coexistence*. Paper for Yale School of the Environment, 2022. Available from the
Northern Rockies Conservation Cooperative, Jackson WY.

Sanborn, Margaret. *The Grand Tetons: The Story of the Men Who Tamed the Western
Wilderness*. New York: G. P. Putnam's Sons, 1978.

Saylor, David J. *Jackson Hole, Wyoming: In the Shadow of the Tetons*. Norman: University of Oklahoma Press, 1971.

Schullery, Paul. *The Bears of Yellowstone*, rev. ed. Worland WY: High Plains Publishing, 1992.

Schullery, Paul. *Lewis and Clark among the Grizzlies: Legend and Legacy in the American West*. Helena MT: Falcon, 2002.

Schullery, Paul. *Yellowstone's Ski Pioneers: Peril and Heroism on the Winter Trail*. Worland WY: High Plains Publishing Company, 1993.

Schwartz, Charles C., and Mark A. Haroldson. "Hazards Affecting Grizzly Bear Survival in the Yellowstone Ecosystem." *Journal of Wildlife Management* 74, no. 4 (2010): 654–67.

Schwartz, Charles C., and Mark A. Haroldson, ed. *Yellowstone Grizzly Bear Investigations: Annual Report of the Interagency Grizzly Bear Study Team, 1999*. Bozeman MT: U.S. Geological Survey, 2000.

Schwartz, Charles C., and Mark A. Haroldson, ed. *Yellowstone Grizzly Bear Investigations: Annual Reports of the Interagency Grizzly Bear Study Team, 2001–2003*. Bozeman MT: U.S. Geological Survey, 2002–2004.

Schwartz, Charles C., Mark A. Haroldson, and Steve Cherry. "Reproductive Performance of Grizzly Bears in the Greater Yellowstone Ecosystem, 1983–2002." *Wildlife Monographs* 161 (2006): 18–24.

Schwartz, Charles C., Mark A. Haroldson, Kerry A. Gunther, and Dave Moody. "Distribution of Grizzly Bears in the Greater Yellowstone Ecosystem, 1990–2000." *Ursus* 13 (2002): 202–13.

Schwartz, Charles C., Mark A. Haroldson, Kerry A. Gunther, and Dave Moody. "Distribution of Grizzly Bears in the Greater Yellowstone Ecosystem, 2004." *Ursus* 17, no. 1 (2004): 63–66.

Schwartz, Charles C., Mark A. Haroldson, and Karrie West, ed. *Yellowstone Grizzly Bear Investigations: Annual Reports of the Interagency Grizzly Bear Study Team 2004–2006*. Bozeman MT: U.S. Geological Survey, 2005–2011.

Schwartz, Charles C., Mark A. Haroldson, and Karrie West, ed. *Yellowstone Grizzly Bear Investigations: Annual Reports of the Interagency Grizzly Bear Study Team 2007*. Bozeman MT: U.S. Geological Survey, 2005–2011.

Schwartz, Charles C., Mark A. Haroldson, and Karrie West, ed. *Yellowstone Grizzly Bear Investigations: Annual Reports of the Interagency Grizzly Bear Study Team 2008*. Bozeman MT: U.S. Geological Survey, 2005–2011.

Schwartz, Charles C., Mark A. Haroldson, and Karrie West, ed. *Yellowstone Grizzly Bear Investigations: Annual Reports of the Interagency Grizzly Bear Study Team 2009*. Bozeman MT: U.S. Geological Survey, 2005–2011.

Schwartz, Charles C., Mark A. Haroldson, and Karrie West, ed. *Yellowstone Grizzly Bear Investigations: Annual Reports of the Interagency Grizzly Bear Study Team 2010*. Bozeman MT: U.S. Geological Survey, 2005–2011.

Schwartz, Charles C., Sterling D. Miller, and Mark A. Haroldson. "Grizzly Bear." In *Wild Mammals of North America: Biology, Management, and Conservation*, edited by George A. Feldhamer, Bruce C. Thompson, and Joseph A. Chapman, 556–86. Baltimore: Johns Hopkins University Press, 2003.

Sellars, Richard West. *Preserving Nature in the National Parks: A History*. New Haven CT: Yale University Press, 1997.

Sellars, Susan Lamadrid. "The Grizzly State of the Endangered Species Act: An Analysis of the ESA's Effectiveness in Conserving the Yellowstone Grizzly Bear Population." *Land and Water Law Review* 29, issue 2 (1994): 469–503.

Smith, Bruce L., and Stanley H. Anderson. "Patterns of Neonatal Mortality of Elk in Northwest Wyoming." *Canadian Journal of Zoology* 74 (1996): 1229–37.

Smith, Bruce L., Katherine C. McFarland, Fred G. Lindzey, and Tom Moore. "Neonatal Mortality of Elk in Areas with and without Grizzly Bears." *University of Wyoming— National Park Service Research Center Annual Report* 23, article 15 (1999): 159–62.

"State Animal, Bird, and Fish Preserves." In *Wyoming Compiled Statutes 1945, Containing the General Laws of Wyoming Annotated*, vol. 3, article 6, p. 664. Indianapolis: The Bobbs-Merrill Company, 1946.

"State Animals, Bird and Fish Preserves, and Blue Holes State Park." In *Session Laws of the State of Wyoming*, passed by the Twenty-ninth State Legislature, chapter 48. Casper WY: Prairie Publishing, 1947.

Steely, James W. *Administrative History, John D. Rockefeller, Jr. Memorial Parkway, Teton County, Wyoming*. May 2022. Organization of American Historians, Denver, Colorado. Grand Teton National Park, Moose WY.

Stephenson, John, and Steven L. Cain. "Wildlife Management: Human-Bear Interface." In *Wildlife Conservation, Management, and Research 2011*, 42–43. Grand Teton National Park, Division of Science and Resource Management, Moose WY.

Stirling, Ian, and Andrew E. Derocher. "Factors Affecting the Evolution and Behavioral Ecology of the Modern Bears." In *Bears, Their Biology and Management* 8 (1990): 189–204.

Stoen, Ole-Gunnar, Andres Ordiz, Alina L. Evans, Timothy G. Laske, Jonas Kindberg, Ole Frobert, Jon E. Swenson, and Jon M. Arnemo. "Physiological Evidence for a Human-Induced Landscape of Fear in Brown Bears." *Elsevier* 152 (2015): 244–48.

Stringham, Stephen F., Lynn L. Rogers, and Ann Bryant. "Have Black and Grizzly Bears Become More Dangerous? Insights from Human-Bear Fatality Trends." *Research Gate*, May 30, 2019, updated January 16, 2020. DOI: 10.13140/ RG.2.2.28390.88648.

Turbak, Gary. "Swapping Conflict for Conservation." *National Wildlife*, February 1, 2004. https://www.nwf.org/donateheader?sc_camp= B75F90F235E54DF2A043295ADB8B51D4.

Turner, Harold Mapes. Oral Interview, August 1, 2017. Jackson Hole Dude Ranching Folklife Project. Utah State University. https://digital.lib.usu.edu/digital/collection /p16944coll34/id/484/rec/1.

U. S. Department of the Interior, Fish and Wildlife Service. *Amendment Listing the Grizzly Bear of the 48 Conterminous States as a Threatened Species*. Federal Register 40, no. 145 (Monday, July 28, 1975): 31734–31736. Washington D C: U.S. Government Printing Office.

U. S. Department of the Interior, Fish and Wildlife Service. *Biological Opinion for Caribou-Targhee National Forest, Targhee National Forest Range Allotments* (May 17, 2016): 01EIFW00–2016-F-0385. Boise: Idaho Fish and Wildlife Office.

U. S. Department of the Interior, Fish and Wildlife Service. *Environmental Conservation Online System*, March 2024. https://ecos.fws.gov/ecp/boxscore.

U. S. Department of the Interior, Fish and Wildlife Service. *Grizzly Bear*. https://www.fws.gov/species/grizzly-bear-ursus-arctos-horribilis.

U.S. Department of the Interior, Fish and Wildlife Service. *Grizzly Bear Recovery Plan*, 1982. State Documents Collection, Montana State Library, Helena.

U.S. Department of the Interior, Fish and Wildlife Service. *Grizzly Bear Recovery Program 2022 Annual Report*, 2023. https://www.fws.gov/sites/default/files/documents/2022%20GBRP%20Annual%20Report.pdf.

U. S. Department of the Interior, Fish and Wildlife Service. *History of Grizzly Bear Recovery, a Timeline*. 2021. https://www.fws.gov/species/grizzly-bear-ursus-arctos-horribilis.

U.S. Department of the Interior, National Park Service. *Environmental Assessment and Finding of No Significant Impact, Flagg Ranch Development Plan, John D Rockefeller, Jr. Memorial Parkway*, 1994. Grand Teton National Park, Moose W Y; and N P S Intermountain Regional Office Library, Denver C O.

U.S. Department of the Interior, National Park Service. *Management Policies*. Washington D C: Government Printing Office, 1988, 2001, 2006. npshistory.com/agency_history.htm#policy.

U.S. Department of the Interior, National Park Service. *Master Plan for Grand Teton National Park*. Moose W Y: Grand Teton National Park, 1976. Also available at nps.gov/grte.

U.S. Department of the Interior, Fish and Wildlife Service. *Proposed Revision of Special Regulations for the Grizzly Bear*. Federal Register 54, no. 199, Tuesday, October 17, 1969, 42524–42528. Washington D C: U.S. Government Printing Office.

U.S. Department of the Interior, Fish and Wildlife Service. *Proposed Rule Designating the Greater Yellowstone Ecosystem Population of Grizzly Bears as a Distinct Population Segment; Removing the Yellowstone Distinct Population Segment of Grizzly Bears from the Federal List of Endangered and Threatened Wildlife*. Federal Register 70, no. 221, Thursday, November 17, 2005, 69854–69884. Washington D C: U.S. Government Printing Office.

U.S. Department of the Interior, Fish and Wildlife Service. *Proposed Rule Removing the Greater Yellowstone Ecosystem Population of Grizzly Bears from the Federal List of Endangered and Threatened Wildlife.* Federal Register 81, no. 48, March 11, 2016, 13174–13227. Washington DC.: National Archives and Records Administration.

U.S. Department of the Interior, National Park Service. *Record of Decision (ROD) for the Moose-Wilson Corridor Final Comprehensive Management Plan/Environmental Impact Statement (Final Plan/EIS)*, 2016. Grand Teton National Park, Moose WY; and NPS Intermountain Regional Office Library, Denver CO.

U.S. Department of the Interior, National Park Service. *Report of the Director, National Park Service to the Secretary of the Interior for the Fiscal Year Ended June 30, 1929, and the Travel Season 1929.* Washington DC: Government Printing Office, 1929. npshistory.com/publications/ annualreports/director/1929.pdf.

U.S. Department of the Interior, National Park Service. *Transportation Plan Environmental Impact Statement, Record of Decision*, 2007. Grand Teton National Park, Moose WY; and NPS Intermountain Regional Office Library, Denver CO.

van Manen, Frank T., Michael R. Ebinger, David D. Gustine, Mark A. Haroldson, Katharine R. Wilmot, and Craig L. Whitman. "Primarily Resident Grizzly Bears Respond to Late-Season Elk Harvest." *Ursus* 30, no. 1 (2019): 1–15.

van Manen, Frank T., Mark A. Haroldson, and Bryn E. Karabensh, ed. *Yellowstone Grizzly Bear Investigations: Annual Reports of the Interagency Grizzly Bear Study Team, 2015–2021.* Bozeman MT: U.S. Geological Survey, 2016–2022.

van Manen, Frank T., Mark A. Haroldson, and Suzanna C. Soileau, ed. *Yellowstone Grizzly Bear Investigations: Annual Reports of the Interagency Grizzly Bear Study Team, 2014.* Bozeman MT: U.S. Geological Survey, 2015.

van Manen, Frank T., Mark A. Haroldson, and Karrie West, ed. *Yellowstone Grizzly Bear Investigations: Annual Reports of the Interagency Grizzly Bear Study Team, 2011–2012.* Bozeman MT: U.S. Geological Survey, 2012–2013.

van Manen, Frank T., Mark A. Haroldson, Karrie West, and Suzanna C. Soileau, ed. *Yellowstone Grizzly Bear Investigations: Annual Reports of the Interagency Grizzly Bear Study Team, 2013.* Bozeman MT: U.S. Geological Survey, 2014.

Webb, Melody. *A Woman in the Great Outdoors: Adventures in the National Park Service.* Albuquerque: University of New Mexico Press, 2003.

White, P. J., Kerry A. Gunther, and Frank T. van Manen, ed. *Yellowstone Grizzly Bears: Ecology and Conservation of an Icon of Wildness.* Bozeman MT: Yellowstone Forever, 2017.

Whittlesey, Lee H., and Sarah Bone. *The History of Mammals in the Greater Yellowstone Ecosystem, 1796–1881.* Seattle: Kindle Direct, 2020.

Wondrak Biel, Alice. *Do (Not) Feed the Bears: The Fitful History of Wildlife and Tourists in Yellowstone.* Lawrence: University of Kansas Press, 2006.

Wyoming Game and Fish Department. "Black Bear Harvest Report." Table 1-H. https://wgfd.wyo.gov/WGFD/media/content/Hunting/2021_BlackBear _HarvestReport.pdf 2021.

Ziesler, Pamela S., and Claire M. Spalding. *National Park Service Visitor Use Statistics, 2021*. Fort Collins CO: NPS Social Science Program, Public Use Statistics Office, February 2022. https://irma.nps.gov/STATS/SSRSReports/park%20Specific%20Reports/Annual%20Park%20Recreation%20Visitation%20Graph%20(1904%20-%20Last%20Calendar%20Year)?Park=GRTE.

Index

hunting of bears, 7–8, 97, 144, 146, 203; of black bears, 158–59; camps for, 52; in Grand Teton NP, 113, 139, 147–49, 152; of grizzly bears, 33, 41, 146–48, *148*, 150–51; historic accounts of, 13, 16–18, 37, 146, 150; in Jackson Hole National Monument, 149; in John D. Rockefeller Parkway, 5, 8, 42, 136, 147–48, 158–59, 164–65; National Academy of Sciences comments on, 151; nuisance hunt proposal, 49, 159–60; permit numbers and costs, 33, 146, 150–51; policy in national parks, 30, 144, 158–59, 164–65; as population estimation tool, 29; post delisting, 163, 165; prehistoric, 9, 85; seasons and zones, 27, 146–47, 149; techniques used, 10, 101, 145, 158; in Yellowstone NP area, 37, 144, 147–48, 151, 159–60, 166
hyperphagia/hyperphagic, 57, 174, 215

Interagency Grizzly Bear Committee (IGBC), 49, 193; management guidelines, 48, 103, 108, 183; Yellowstone Ecosystem Subcommittee of, 49, 52, 103, 162
Interagency Grizzly Bear Study Team, 1, 43, 46, 58, 89, 145, 169, 199, 212; employees of, 79, 84, 202, 213

Jackson, William Henry, 13
Jackson Hole Hereford Ranch, 98, 111. *See also* Porter, Robert Bruce, and Gill family (ranchers)
Jackson Hole National Monument, 25, *26*, 27–28, 149
Jackson Hole Wildlife Foundation, 193, 197–98
Jackson Lake, 2, 61, 86, 89, 138, 144, 149, 167; bear death at, 151; boating and fishing on, 5, 146, 200; grizzly bear

distribution around, 2, 46, 48, 50
Jackson Lake Dam, 9, 23, 81, 95, 130t, 137
Jackson Lake Lodge: bears at, 46, 69, 116, 131t, 139, 202; clinic, 137; junction, 24, 99; overlook, 194; river concession, 135
Jenny Lake (Jennie's Lake), 59, 172; bear-caused human injuries at, 130t, 134, 137; bear hunting at, 147; bears at, 13, 28, 36, 51, 69, 183, 194; campground, 71, 99, 137, 183, 195; dam proposed at, 23
John D. Rockefeller, Jr. Memorial Parkway, 2, 5, 77, 81, 89, 144, 153, 161, 167, 220; bear management in, 3, 40, 48, 123, 162, 166, 209; bears killed, moved into, or trapped in, 46, 82, 159, 216; boundaries of, *6*; critical habitat in, 47; dump in, 68; establishment of, 154–55; grizzly bear recovery in, 48, 159, 191; hunting in, 147, *148*, 160, 164–65. *See also* Flagg Ranch; hunting of bears
Johnson, Lady Bird, 61–64
Johnson, William, 145
Jonkel, James "Jamie" (biologist), 215, 217–18
JY Ranch, 15, 61, 63–64, 85, 86; plans for transfer of, 3, 185

Kelly, dump, 69, 71; livestock grazing in/near, 95, 99, 113, 198
Kerr, Robert "Bob" (park superintendent), 43, 47, 50
Kilpatrick, Steve (biologist), 196–97
Knight, Richard "Dick" (biologist), 84–85, 91, 93, 103

Langford, Nathaniel P., 11–12
Laurance S. Rockefeller Preserve, 15, 79, 186
Lawrence, Slim, 40, 145, 148, 149
Leigh, Beaver Dick, 11
Leigh Lake, 11, 51

to, 2, 25, 48, 95, 155, 158–59, 161, 215–17; proposed expansion of, 23, 25, 86, 154; research/controversy in, 4, 37–38, 40, 45, 53, 61, 151, 168–69, 182; thermal features and protection of, 154–56. *See also* abundance and distribution of bears; aversive conditioning; bears as pets or for sale; bears/bear viewing as visitor attraction; feeding of bears; garbage dumps; hazing of bears; hunting of bears; National Park Service (NPS): management of bears in Yellowstone; Management Situations (MS) for grizzly bears; predation; trapping/traps of or for bears

www.ingramcontent.com/pod-product-compliance
Lightning Source LLC
Chambersburg PA
CBHW030940100225
21630CB00004B/12

* 9 7 8 1 4 9 6 2 3 6 2 7 2 *